Transgenic Plants in Agriculture

Ten Years Experience of the French Biomolecular Engineering Commission

Under the direction
of Professor Axel Kahn
Ex-President of the French Biomolecular Engineering Commission

Translated into English by ALPHATRAD
with the helpful scientific advice of Dr. Ian D. Small

Ministère de l'Agriculture
et de la Pêche
DGAL
251, rue de Vaugirard
75732 Paris Cedex 15

Ministère
de l'Environnement
DPPR
20, avenue de Ségur
75230 Paris

John Libbey
EUROTEXT

Transgenic Plants in Agriculture

Ten Years Experience of the French Biomolecular Engineering Commission

Éditions John Libbey Eurotext
127, avenue de la République, 92120 Montrouge, France
Tél : 01.46.73.06.60

John Libbey and Company Ltd
13, Smiths Yard, Summerley Street,
London SW18 4HR, England
Tel : (1) 947.27.77

John Libbey CIC
Via L. Spallanzani, 11, 00161, Rome, Italy
Tel : (06) 862.289

ISBN 2-7420-0201-4

Cover pictures : J. Weber, INRA

Contents

Authors List

Aigle M., ex-Vice-President of the CGB, Professor, Université de Bordeaux, 1, rue Camille-Saint-Saens, 33077 Bordeaux Cedex, France ; Tel : 05 56 99 90 17 ; Fax : 05 56 99 42 99.

Béranger S., ex-Head of the Unit for Regulatory Affairs, Ministry of Agriculture, Fisheries and Food, 251, rue de Vaugirard, 75732 Paris Cedex 15, France ; Tel : 01 49 55 55 84 ; Fax : 01 49 55 59 48.

Casse F., Member of the CGB, Professor, Laboratoire de Biochimie et Physiologie Végétale, 9, place Viala, 34060 Montpellier Cedex 01, France ; Tel : 04.99.61.26.14 ; Fax : 04.67.52.57.37.

Chaufaux J., Research engineer, INRA, Domaine de la Minière, Station de Recherches de Lutte Biologique, 78280 Guyancourt, France ; Tel : 01.30.83.36.32 ; Fax : 01.30.43.80.97.

Chupeau Y., Member of the CGB, Senior Scientist, INRA, Laboratoire de Biologie Cellulaire, Route de Saint-Cyr, 78026 Versailles Cedex, France ; Tel : 01 30 83 30 16 ; Fax : 01 30 83 30 99.

Kahn A., ex-President of the CGB, Senior Scientist, INSERM, ICGM, CHU Cochin, INSERM U 129, 24, rue du Faubourg-Saint-Jacques, 75014 Paris, France ; Tel : 01 44 41 24 24 ; Fax : 01 44 41 24 21.

Laude H., Member of the CGB, Senior Scientist, INRA, Domaine de Vilvert, 78352 Jouy-en-Josas, France ; Tel: 01.34.65.26.13 ; Fax : 01.34.65.26.21

Le Buanec B., Secretary General to the International Seed Federation, 7, chemin du Reposoir, 1260 Nyon, Switzerland ; Tel : 00 41 22 361 99 77 ; Fax : 00 41 22 361 92 19.

Lereclus D., Senior Scientist, Institut Pasteur, Unité de Biochimie Microbienne, 28, rue du Docteur-Roux, 75724 Paris Cedex 15, France ; Tel : 01 45 68 88 13 ; Fax : 01 45 68 89 38.

Pascal G., Member of the CGB, Senior Scientist, Head of the Centre National d'Études et de Recommandations sur la Nutrition et l'Alimentation (CNERNA,

CNRS), 16, rue Claude-Bernard, 75231 Paris Cedex 05, France ; Tel : 01 42 75 93 24 ; Fax : 01 44 08 72 76.

Reverbori H., Head of the Unit for Regulatory Affairs, Ministry of Agriculture and Fisheries, 251, rue de Vaugirard, 75732 Paris Cedex 15 ; Tel : 01 49 55 55 84 ; Fax : 01 49 55 59 48.

Sanchis V., Senior Scientist, Institut Pasteur, Unité de Biochimie Microbienne, 28, rue du Docteur-Roux, 75724 Paris Cedex 15, France ; Tel : 01 45 68 88 13 ; Fax : 01 45 68 89 38.

Schoonejans E., Unit for Regulatory Affairs, Ministry of Agriculture and Fisheries, 251, rue de Vaugirard, 75732 Paris Cedex 15 ; France ; Tel : 01 49 55 58 86 ; Fax : 01 49 55 59 48.

Thuriaux P., ex-Member of the CGB, Senior Scientist, Commissariat à l'Énergie Atomique Saclay, Service de Biochimie et Génétique Moléculaire, Bâtiment 142, 91191 Gif-sur-Yvette, France ; Tel : 01.69.08.35.86 ; Fax : 01.69.08.47.12

Introduction

Risk Assessment and the Deliberate Release of Transgenic Plants: the French Experience

Axel Kahn*

Summary

Biotechnology may be defined as the use of living cells to produce different substances. The range of possibilities opened up by biotechnology has been greatly extended by the development of genetic engineering. The precise transfer of a gene from any living cell to a plant can now replace the conventional methods of selection used to obtain new varieties with new phenotypic characteristics. Genetic engineering is certainly not dangerous in itself and it allows plant geneticists to do things which would have been impossible otherwise.

For this reason, a case-by-case assessment of the potential risks associated with the large-scale use of such transgenic plants seems entirely justified. In France, the Biomolecular Engineering Commission (BEC) (in French, Commission du Génie Biomoléculaire, CGB) is the adviser to the competent authorities that authorize the deliberate release and commercialization of genetically modified organisms, especially transgenic plants.

Since 1986, this commission has authorized more than 450 field tests on transgenic plants, on more than 3,000 different release sites. Consequently, they have more experience in this area than anyone else in Europe. This experience has allowed the CGB to develop a "philosophy" whose main characteristics can be summed up as follows:

* Ex-President of the Biomolecular Engineering Commission, Paris, France.

1. The assessment of the potential risks associated with a transgenic plant requires the precise characterization of the plant, of the transgene actually integrated and of the behaviour of the transgenic plant in its natural ecosystem.
2. Certain phenomena that may not be detected in small-scale tests may be noted in the large-scale cultivation of transgenic plants.
3. In the realm of transgenic plants, any identifiable risk would probably be considered unacceptable to Europeans.
4. The goal of genetic engineering in agriculture is not only to produce more, but also to produce more safely.

Biotechnology can be defined as the use of living cells to produce food and non-food substances. It is extremely ancient as it includes most agriculture, the production of cheese, wine, beer, etc. However, genetic engineering is more recent. It involves all of the techniques that, benefiting from the universal nature of the genetic code, can be used to program living organisms to carry out the genetic instructions contained in one or several genes obtained from another organism. The use of genetic engineering methods by biotechnology has greatly expanded the possibilities, especially as regards the improvement of plants.

The principle for the selection of cultivated plants, empirically carried out for the last 10,000 years, is to cross males and females of a given species. The plants inheriting the desired characteristics are chosen from among the descendants. Genetic engineering can now obtain the same results by directly identifying the genes responsible for the desired character and transferring them to a variety of plants. The main difference between classical selection and improvement by genetic engineering is that the latter crosses the species barrier, so that a gene for improvement can be chosen from almost any living species (micro-organisms, plants or animals). In addition, the transgenic approach limits the genetic modification to the characteristic studied, whereas the selection of the character using traditional methods is accompanied by the transmission of many other hereditary characters that by chance are linked to it. For this reason, genetic engineering provides much better control for obtaining new varieties than selection after cross-fertilization of parental species. Hence this method is not inherently dangerous in itself.

The great diversity of possible transgenes (the genes obtained from different organisms and transferred to the plant) may be used to create many new plant varieties. In certain cases, genetic engineering may facilitate farming practices and increase the yield: for example, transfer of herbicide, insect or disease-resistance genes; yield improvement in difficult climatic conditions or poor soils; reduced need for fertilizers and creation of hybrid species. In other cases, it may help to improve the quality of the plant, for example by enriching it in essential amino acids or modifying the level of certain food components. The modifications may also help improve the industrial use of plants by increasing the biofuel content or by changing the quantity and quality of the oils used in chemistry or cosmetics. In the future, plants may be used to produce substances that have never been found within the world of agriculture: different enzymes, other types of proteins for drugs, plastics, etc.

These examples indicate that genetic engineering goes far beyond the possibilities of nature. For this reason, concern with safety is justified on a case-by-case basis. It is easy to assess that a transgene is not dangerous for animal or human health. For example, the safety is obvious when the transgene comes from a plant that is ordinarily consumed. However, it is necessary to more carefully investigate the absence of a change in behaviour of a transgenic plant that may allow it to spread like a weed in crops. For this reason, at international level, the first tests on transgenic plants were surrounded by a series of precautions and were carried out under conditions where these factors could be tested. In the United States, over 3,000 tests on transgenic plants have been carried out in the field after receiving the necessary approval.

In France, in accordance with the criteria noted in European directive 90-220, the deliberate release of genetically modified organisms (GMOs) is studied by the Biomolecular Engineering Commission (in French: CGB), created in 1986 and slightly modified in 1992, after the law of July 13th of that year incorporated the European directives into French law.

The people creating this commission immediately wanted to open it up to feelings that go far beyond the strict scientific examination of the projects assessed. For this reason, the CGB includes representatives from industry, unions and workers, consumers movements and environment protection groups. The Parliament, the Department of Health and the Department of Research are also represented. Slightly over half of the members of the CGB are scientists with the skills required for the technical understanding of the projects proposed. Since 1987, when the CGB began to assess the potential risks associated with the release of GMOs, it has already approved 450 proposals, most of them involving transgenic plants, used in tests on over 3,000 sites. This activity places the CGB in second position (after the United-States) in the deliberate release of GMOs. The experience acquired by the CGB since 1986 has allowed it to develop a "philosophy" concerning the use of genetic engineering. This can be summed up as follows:
- the goal of modern science and genetic engineering is not only to produce more but also to produce with increased safety;
- it is possible to obtain an exact assessment of the potential risks involved in the release of a GMO in the environment when the modification is as limited as possible, its nature perfectly known and when the past research has eliminated any detectable risk;

- the acceptability of the risk is relative and depends on the importance attached by the population to the goals set by the researchers. Risks are accepted in medicine if an attempt is made to cure a disease that is either incurable or fatal. In agriculture, the population in developed countries probably considers any identifiable risk as unacceptable as they have not expressed any specific needs or requests in this field;
- the assessment of the risk is sequential and should systematically refer to the definition of the word "risk", that the degree of danger multiplied by the probability. There is obviously no risk involved if there is no danger inherent in an experiment. However, there is also no risk if the theoretical consequences of an experiment are very dangerous but there is zero probability. The CGB studies the potential danger in the light of all the knowledge acquired concerning the transgene, its interactions with the plant in which it will be introduced and past tests. The Commission estimates the probability of a theoretical danger, once it is identified. Release is authorized only when all identifiable risks are eliminated.

In addition to the assessment of the risk resulting in the authorization or refusal of a test, the CGB feels that it is best for giving long-term consideration to the ecological, agricultural and economic consequences of certain practices related to the commercialization of GMOs, mainly transgenic seeds. The Commission follows the files, from the first tests, often carried out on a few square meters, right until marketing approval has been requested. These files often evolve *en route* as a result of the CGB recommendations. Consequently, the French Commission feels that it is its duty to draw attention to any uncertainty related to commercialization as soon as possible. These uncertainties may not necessarily correspond to an identified or potential risk but also to unknown factors involved in the new practices that may have socioeconomic consequences. For this reason, the French position concerning herbicide-resistant plants differs from that of similar European agencies. This is illustrated by the recent example of a request to commercialize transgenic rape with a herbicide resistance gene. The CGB first met with the different partners (plant molecular geneticists, manufacturers, population geneticists, weed specialists) in order to determine whether the escape of the transgenic rape *via* pollen and seeds was likely, and whether the transgene could be transferred to other interfertile species, and if so, what was the probability of this occurring. Contrary to the assurances initially given by the manufacturers and other European agencies, it turned out that the spread of transgenic rape was a certainty and that the spread of the transgene in inter-specific hybrids was possible. However, there are uncertainties about the probability that such species can stably settle and create an ecological problem.

Based on this analysis, the CGB distinguished two situations. In certain cases, the herbicide at which the herbicide resistance gene is aimed has no major value and may easily, technically and economically, be replaced by similar products at the same price. In this case, the risk is mainly industrial and is taken by the company requesting the authorization to market transgenic seeds, since reduced efficacy of the herbicide (due to generalized resistance following escape of the transgene) would lead to a loss of interest in the transgenic plant. In other cases, the resistance genes concern herbicides without an equivalent product. The loss of effectiveness would then greatly penalize farmers or even raise major economic problems for European agriculture. The CGB considers that it has a duty to point this out to all those involved in agriculture and the competent authorities. No short-term disaster resulting from the cultivation of such plants is anticipated as demonstrated by many preliminary field tests. In fact, in certain sites, the dissemination of the resistance gene towards adventitious species may be suppressed by eradicating undesirable plants in using a mixture of existing herbicides. However, this solution requires biological monitoring to detect such unfavourable evolution when it occurs, for example the appearance of many adventitious species that resist one or more herbicides that are of great value and are used extensively. The French Commission asks applicants to provide a detailed outline of the measures that they will use to detect such phenomena and to eliminate them. The CGB has also recommended that the first authorizations for herbicide-tolerant rape be temporary and conditional and be accompanied by a system of biological monitoring to determine the factors indicative of different possible scenarios following the widespread use of transgenic seeds. A commission representing all parties would be set up to deal with these factors and give its opinion. This will be used to confirm or withdraw the temporary authorization.

There is a great deal of potential for improving varieties using transformation in developed countries as well as, in principle, in developing countries. The former can use it to promote more ecological farming (by reducing the need for pesticides and different plant protection products or even fertilizers in the future), more competitive farming, to improve the quality of life of farmers and to develop new markets. In this way, the prospects opened up by non "food" farming, whether for biofuels, chemicals, cosmetics or drugs, are undoubtedly attractive in an era when the state of the farm markets, within the context of the common agricultural policy in Europe, leads to the expansion of land left fallow and the progressive abandonment of farming in unfavourable zones. For the developing countries, the arrival of pest-resistant species that adapt to arid conditions with an increased food value will certainly be beneficial. However, the economic conditions must let poor countries take advantage of these improvements. It should not turn into another domain where the Third World depends on the industrialized countries. In our countries, the success of these techniques will depend on their economic value (which will prompt farmers to use them) and their acceptance by consumers that have been informed of what has been done, of the tests carried out in order to determine the safety of the approach, and of the goals involved.

1

Economic Prospects of Transgenic Plants

B. Le Buanec*

Summary

One of the greatest challenges that the world will have to face over the next 20 years is how to feed an increasing population within the framework of an environmentally friendly agriculture.
The use of improved plant varieties is one way to achieve this goal. The vehicle for improved varieties is high quality seed. The world seed trade currently amounts to about $ 30 million (US).
Plant breeding has played an important role in the development of agriculture since the beginning of the century and should continue to play a basic role in the years to come. The annual plant breeding budget in developed countries can be estimated at $ 800 to 900 million (US).
Plant breeding has integrated technical and scientific advances, a sign of efficiency, and now routinely uses biotechnology as a tool. Plant biotechnology research budgets can be estimated at $ 2 billion (US) or 12 billion French francs.
Transgenic varieties of various crops, resistant to insects, viruses and herbicides and with improved agricultural traits are currently mainly marketed in North America but also in China, Argentina. Three million hectares of transgenic plants were cultivated in 1996 and 13 million in 1997. This should exceed 30 million hectares in 1998, except in Europe where transgenic plants are poorly established.

* Secretary General to the International Seed Federation, Nyon, Switzerland.

Feeding the World

In 1996, the world grain reserves decreased to 261 million metric tons, corresponding to a security index of 51, which is the lowest ever obtained (*Table I*). This resulted in a very strong increase in grain prices on the world seed market leading to serious difficulties for low-income food-deficit countries (LIFDCs), who had to cope with a 20% increase of their cereal imports bill within a year. It is clear that the food situation in the world is still vulnerable and contrary to some ideas maintaining seed production volume remains an important objective.

Table I. Evolution of grain reserves[1].

Year	Grain Reserves (MT)	Safety Index[2]
1960	199	103
1970	165	75
1980	191	56
1989	222	54
1995	305	59
1996	262	51
1997[3]	292	55
1998[3]	302	57

1. Compilation of different sources, from the FAO.
2. Days supply.
3. Estimates.

Estimates for 1997 and previsions for 1998 seem to indicate progress but are still worrying considering that according to the FAO, the reasonable level is 65. This will have serious consequences for low-income food deficit countries (LIFDC's).

In India, 41.5 million hectares of rice are grown. The average yield is slightly higher than 2.8 t/ha, providing an annual production of about 117 million metric tons. For the last few years, the yield has stagnated, especially in high production zones. This will no longer provide a 3% increase in production per year, the amount required to feed the population increase.

Contrary to the current tendency to reach a ceiling in food production, the world population tends to increase on a regular basis, as indicated in *Table II* [1].

The average yearly increase in the world population has been predicted to be 1.6% between 1995 and 2000, 1.5% between 2000 and 2005, 1.2% between 2010 and 2015 and 1% between 2020 and 2025 (*Table III*). The average increase in population, currently about 90 million per year, should drop to about 83 million between 2020 and 2025, which is approximately the 1985 level.

Table II. World population (in billions).

Year	Population
1950	2.5
1990	5.3
2000	6.3
2025	8.5

Table III. Average annual population increase (%).

	1985-1990	2020-2025
The most developed regions	0.5	0.2
The least developed regions	2.1	1.2
The least developed countries	2.8	1.7
Other countries	2.0	1.0

The lowest United Nations estimate of the world population in 2025 is 7.3 billion inhabitants, almost 2 billion more than today. The upper estimate predicts a population of 9.4 billion.

The highest increases are currently occurring in West, East and Central Africa.

However, *per capita* food production is stagnating in most regions with a high population increase or is even decreasing as is true in Africa, as indicated in *Figure 1* [2].

In addition to these unfavourable quantitative aspects, quality has to be taken into account. Joseph Klatzmann's demonstration [2] is highly informative. It would be necessary to produce 50% more than we do today to ensure a satisfactory food supply for the current world population. Current farm production would need to be multiplied by 3.5 in order to feed ten billion inhabitants (expected by 2050 according to the average United Nations estimates).

Since the land under cultivation (about 15 million km^2 today) will undoubtedly decrease due to the increase in population, the only solution for meeting the challenge of feeding the world population is to increase the productivity per unit cultivated. This increase depends on a great many factors such as irrigation, the use of fertilizers, herbicides and pesticides. The use of improved plant varieties is one of the major factors and is often cited as being responsible for 50% of the increase in productivity over the last 30 years.

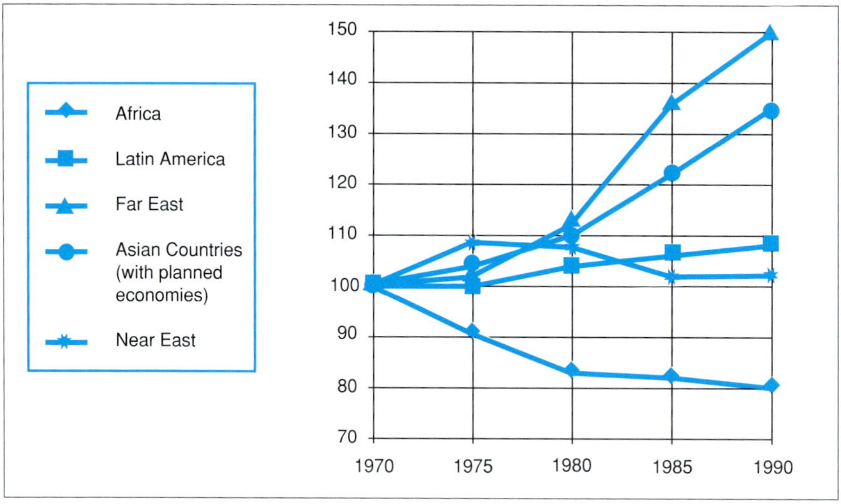

Figure 1. ***Per capita*** **change in food production between 1979 and 1990 [2]**
(1970 index = 100).

The Challenge of Developed Agriculture

While the major challenge confronting the world remains the increase in food production, countries with highly developed agriculture, mainly in Western Europe and North America, are faced with overproduction and agricultural surpluses are inciting governments to reduce the area under cultivation by means of subventions. This does not mean that the agriculture in these countries does not have major challenges to face.

Intensified agriculture in Western Europe and North America made these regions self-sufficient. They then became exporters with the resulting benefits for their economies and world food levels. This has sometimes been accompanied by a pollution of the soil, the water table and surface waters due to the intensive use of fertilizers and pesticides. This has to be corrected without overly reducing productivity. We have seen that a reduction in supplies immediately creates some tension in world markets and affects prices. This has a negative effect on the food supplies in developing countries or on their trade balance or, most often, on both. Hence the idea that developed countries can reduce production, stop exporting and simply return to being self-sufficient is unacceptable both for their economies and the world food supply, at least until the developing countries have also become self-sufficient.

Another important characteristic of the agriculture of developed countries is its integration in a complex agro-food chain with a major impact on employment and the economy. Currently, the share of industrially transformed or prepared foodstuffs, compared with the consumption of unrefined products is 70% in Europe, 90% in the United States and 10% to 20% in many developing countries [3]. Pierre Feillet and J. Rajnchapel-Messaï [4] provide data concerning the added value of the agro-food industries in the agro-food chain (*Table IV*).

Table IV. Added value of the agro-food industries in the agro-food chain (% of the total).

Great Britain, Germany	70-75
France, Belgium, Luxembourg	55
Netherlands, Denmark, Spain, Italy	40-55
Portugal, Ireland	35-39
Greece	15

In France, 1,850,000 people are active in agriculture (including forestry and fishing). This amounts to 7.5% of the total work force. Meat and dairy industries employ about 180,000 people and other food industries about 430,000.

For socioeconomic reasons, it is necessary to stabilize this sector. One way is to better adapt agricultural products to industrial processing. Plant improvement is a very powerful tool for this.

Finally, although agriculture is often only thought of as food production, this is not the case. Twenty per cent of the agricultural production in OECD countries is used for non-food purposes [5]. It is important that agriculture continues to diversify by producing new raw materials. Here too, plant breeding can and should play a major role.

The Seed Industry and Plant Breeding

The world is confronted with the need to significantly increase food production over the coming decades. This increase should occur within the framework of sustainable development and be environment-friendly. On a regional level, particularly in Western Europe and North America, specific questions are being raised concerning the increased respect of the environment, the increased adaptation of agricultural products to the needs of industry and the search for new outlets.

Providing seeds of improved plant varieties to farmers is the best way to make progress in reaching these goals.

It is difficult to assess the world market for seeds and plants for a great many technical and statistical reasons [6]. With certain reservations, it is possible to assess the total world seed consumption at over $ 50 billion (US). Commercial trade accounts for about 30 billion, including twenty billion in countries with a market economy.

Table V shows the value of the seed markets in countries with a significant agricultural sector.

Table V. Value of the seed markets in some countries with an important agricultural sector.

Country	Domestic trade (in billions of US $)	Country	Domestic trade (in billions of US $)
United States	4,500	Great Britain	570
Russia	2,500	Spain	550
Japan	2,500	Poland	400
China	2,500	Hungary	400
France	1,800	Canada	350
Germany	1,500	Netherlands	300
Brazil	1,200	Australia	280
India	900	Austria	170
Argentina	800	Morocco	160
Italy	700	Egypt	140
Total 22,220			

These twenty countries account for over two-thirds of the world seed market.

In addition to these domestic markets, international trade accounts for $ 2.9 billion (US) including 3.3 billions for agricultural crops and 2.3 billion for vegetables and ornamental plants (*Table VI*).

This accounts for about 16% of the world seed market. This general percentage differs according to whether seeds for agricultural crops or seeds for market gardening are involved: 9% of the seeds for agricultural crops, 33% of the seeds for market gardening.

Table VI. World seed export (millions US $).

Maize	532
Forage	427
Potatoes	400
Beets	308
Wheat	75
Other agricultural seeds	590
Horticultural seeds	1,115

There are major differences between the different species and groups of species concerning the seeds for agricultural crops.

The major world products are vegetables and flowers, fodder and beets. The "regional" products are maize and potatoes. The "local" products are seeds and large-seed legumes.

The evolution in trade is provided in *Tables VII* and *VIII*.

Table VII. Evolution in world trade.

	1970	1977	1980	1985	1994	1996
Trade in millions of US $	860	1,076	1,200	1,300	2,900	3,300
1970 base: 100	100	125	140	151	337	283
1977 base: 100		100	111	121	269	207

Table VIII. Evolution in EEC trade (12 countries).

	1977	1980	1985	1993
Value in millions of US $	182	234	346	494
1977 base: 100	100	128	190	271

The international seed trade is increasing steadily and is more important than the main domestic markets. This is especially true of the European Union where the domestic market has stagnated or even declined over the last three years.

The first six exporting countries with a turnover exceeding $ 100 million (US) account for 74% of the world's exports, 86% of the exports of seeds for large-scale farming and 51% of the vegetable seeds (*Table IX*).

It is necessary to bring the world regulations into line in order to continue these international exchanges.

The world turnover in seeds is very low compared with the value of crop production and the production of the agro-food industry. The following estimates indicate the different stages in the chain (*Table X*).

These global world estimates agree with the estimates of the French data provided by P. Feillet *et al.* [4] which also include animal production (*Table XI*).

Table IX. Turnover of the main exporting countries (millions of US $).

	Agricultural crops	Vegetables and flowers	Total
United States	500	200	700
Netherlands	420	200	620
France	432	100	532
Germany	178	35	213
Denmark	116	20	136
Italy	73	30	103
Other	611	530	1,141
Total	2,330	1,115	3,445
% of the first six	74	52	67

Table X. Global world estimates.

Stage	Turnover (billions of US $)
Seeds	30
Plant production	400-500
Processed products	1000-1200

Table XI. Global French estimates.

Stage	Turnover (billions of US $)
Seeds	9
Animal and plant production	330
Processed products	650

The research budget for plant production, the basis of all the world agricultural activity, is relatively difficult to assess. There are few publications on this subject. In 1989, F. Desprez and P. Devaux [7] have estimated it at 4 billion French francs for an annual seed turnover of 90 billion. In 1988, P.B. Joly and C. Ducos [8] estimated it at $ 600 million (US) or about 3.6 billion French francs, for a turnover of $ 15 billion. In both examples, the research budget as a percentage of the turnover is about the same, 4.4% and 4% respectively. Although probably a little low, these figures seem to be coherent. For a commercial seed turnover of $ 20 billion in the developed countries, we can consider that the research budgets devoted to plant breeding by public and private agencies amount to about $ 800 to 900 million.

This global value is of interest in itself. However, it does not take into account the evolution of these budgets over the last twenty years. Although the figures are not readily available, these budgets have increased greatly during the fifties to seventies. They remained stable in the eighties and probably decreased in the nineties.

The following examples, representative of the general situation, support this position (*Figures 2* and *3*).

This stagnation in the research budgets devoted to plant breeding possibly accounts for the slower increase in agricultural production over the last few years. The graph in *Figure 4* [11] clearly shows this trend.

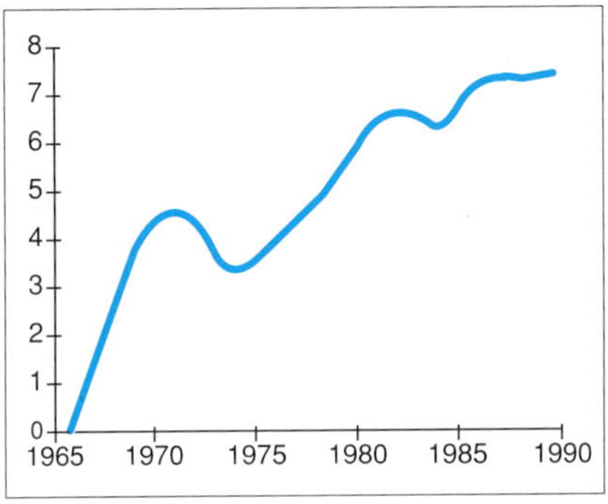

Figure 2. Evolution of the Limagrain research budget from 1966 to 1990 (% turnover) [9].

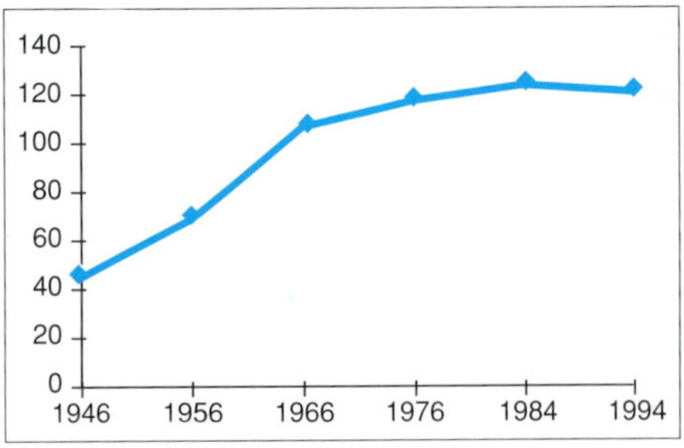

Figure 3. Evolution of the number of scientists employed by the INRA department of plant breeding [10].

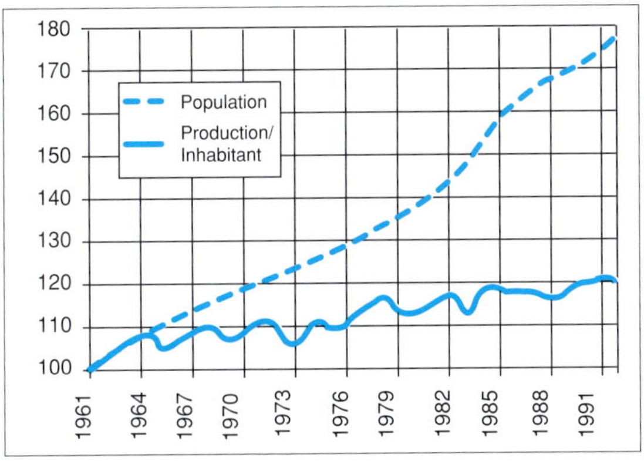

Figure 4. Global evolution of farm production and population between 1961 and 1992 (1961 index: 100) [11].

The Development of Plant Biotechnology

Plant breeding was remarkably ineffective until the beginning of this century if we exclude the period of domestication that probably lasted for several thousand years. The wheat yields during the Roman empire and at the beginning of the 20th Century were about the same (0.8 to 1 metric ton/ha).

In the 20th century, an understanding of Mendelian genetics and the gradual assimilation of scientific and technical discoveries have helped to improve plants and to make considerable progress (*Figure 5*) [12].

The search for variability is a constant goal for the breeder. The evolution of the techniques in this area may be summed up as follows, in chronological order:
- cross-pollination,
- intraspecific hybridization,
- inter-specific hybridization (same ploidy),
- di- or polyploidization,
- embryo rescue,
- fusion of protoplasts,
- recombinant DNA.

If the breeder is prevented from employing genetic engineering (recombinant DNA) in plant breeding, there is a risk of maintaining the slow down observed in the improvement of varieties. This may prevent world agriculture from meeting the challenges it is faced with. As regards the developing countries, G. Persley [13] indicates: "*We have now obtained the same level of productivity in the tropics, at least in the test plots, as in the temperate zones. Little by little, the developing countries are establishing the infrastructure for a much higher farm productivity... Unless highly significant progress is made in productivity (more than currently found in the data provided by the international research centres and in other centres involved in research on tropical farming), I do not feel that we will be able to contradict Malthus' position by the years 2000 or 2030. Significant new progress has to come from genetic engineering.*"

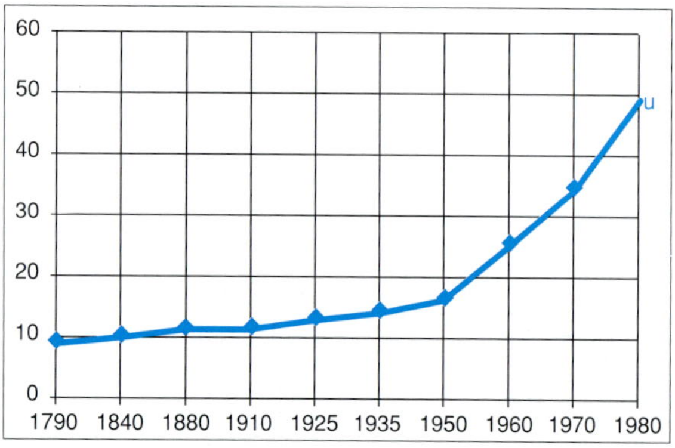

Figure 5. Evolution of the mean yield in grains in France from 1790 to 1980 (Yield q/ha) [12].

It is obvious, as indicated in the OECD report "Biotechnology, Agriculture and Food" [5] that "*even if it were possible to feed an increasing world population, at least until a certain date next century, using 'better practice' technologies as well as political and institutional adjustments on a world scale, the biotechnologies provide a great many additional options to reach the goal of an adequate food supply, at a lower cost, including environmental costs*".

In the developed countries, the huge potential of biotechnology will probably be used to reduce the use of chemicals and better adapt the products for processing rather than to increase yield. However, the distinction is often artificial.

In 1986, the Office for Technology Assessment of the United States Congress (OTA) tried to assess the impact of the new technologies on the growth of agriculture in the United States (*Table XII*). Biotechnology accounts for a major share of the impact of the new technologies in this study.

Table XII. Past and future annual rate of growth in farm production in the United States (%).

	1962-1982	1982-2000		
		Assuming "No new technologies"	Most likely case	Assuming "More new technologies"
Maize	2.6	0.5	1.2	1.6
Cotton	0.1	0.3	0.7	1.0
Rice	1.2	0.2	0.9	1.4
Soybean	1.2	0.8	1.2	1.2
Wheat	1.6	0.7	1.2	1.4

Governments and private companies are convinced of the need to invest in the plant biotechnologies and have set aside huge budgets for this. P.B. Joly and C. Ducos [8] have calculated this investment for 1988. Since this date, investments have considerably increased [14] (*Table XIII*).

Table XIII. Estimated plant biotechnology research budget (millions of US $).

	1988		1992		1996		
	Private	Public	Private	Public	Private	Public	Total
United States	110	30	595	188	960	300	1,260
Other	90	70	–	–	500	180	680
Total	200	100	–	–	1,460	480	1,940

Although it is difficult to determine the share devoted to transgenic plants, these investments have borne fruit. At the end of 1997, field experiments were authorized in 45 countries (*Table XIV*) and 60 species are concerned (*Table XV*) according to C. James [15].

Table XIV. List of 45 countries where field experiments were authorized 1986 to 1997.

South Africa	Chile	Finland	Malaysia	Slovakia
Germany	China	France	Mexico	Sweden
Argentina	Costa Rica	Georgia	New Zealand	Switzerland
Australia	Cuba	Great Brtain	Norway	Thailand
Belgium	Czech Republic	Guatemala	Netherland	Turkey
Belize	Denmark	Hungary	Poland	Ukraine
Bolivia	Egypt	India	Portugal	Uzbekistan
Bulgaria	Spain	Italy	Roumania	Yugoslavia
Canada	United States	Japan	Russia	Zimbabwe

Table XV. List of 60 transgenic crops in field experiments worldwide 1986 to 1997.

Large n° of trials (commercialized) > 150 trials	**Medium n° of field trials (pre-commercialized) 25-150 trials**	**Low n° of field trials (experimental) 1-25 trials**	
Canola/napus	Alfalfa	*Amelachier laevis*	Grape
Canola/rape	Beet	Apple	Kiwi
Cotton	Cantaloupe	*Arabidopsis thaliana*	Lettuce
Maize com	Carnation	Asparagus	Lupins
Melon	Flax	Barley	Mustard, Brown
Potato	Gourd	Belladona	Mustard, Indian
Soybean	Rice	Birch	Papaya
Tobacco	Sunflower	Blueberry	Peanuts
Tomato		Brocolli	Peas
		Cabbage	Petunia
		Carrot	Pepper
		Cauliflower	Plum
		Chicory	Poplar
		Chrysanthemum	Raspberry
		Clover	Serviceberry
		Cranberry	Spruce
		Creeping bent grass	Strawberry
		Cucumber	Sugar cane
		Eggplant	Sweet potato
		Eucalyptus	Wheat
		Gerbera	Walnut
		Gladiolus	

The research is mainly carried out on the resistance to viruses, fungal diseases, insects, herbicides and on the improvement of product quality. The joint report published by the Academy of Sciences and the CADAS [16] as well as the article by V.C. Knauf [17] provide interesting details concerning the possibilities of transgenic plant breeding.

Cultivated areas with transgenic plants are being developed throughout the world as these varieties seem appreciated by farmers that can choose between transgenic and non-transgenic varieties. This evolution is exceptional considering the speed of adoption of other forms of technical progress in agriculture. *Table XVI*, according to C. James [15] shows this evolution.

Table XVI. Cultivated areas with transgenic plants in the world.

Species	**Countries**	**Surfaces in Million ha**		
		1996	**1997**	**1998**
Soybean	United States	0.5	4.45	13.0
	Argentina		1.0	2.0
Maize	United States	0.3	3.2	8.0
	Europe			0.1
	Argentina			0.1
Cotton	United-States	0.8	1.32	2.0
	Australia		0.2	0.3
Rape	United-States	0.1	0.12	0.4
	Canada		1.0	2.0
Tobacco	China	1	1.0	1.0
Vegetables (including potatoes)	United-States	0.2	0.44	0.5
Total		**2.9**	**12.73**	**29.4**

It is interesting to note that the research on transgenic plants started at the same period at the beginning of the eighties in Europe and United States and field trials were made at the same time in 1987 in Europe and the United States. However, none or few of the practical applications of transgenic plants have been developed in Europe due to the specific political context in this region.

Conclusions

Although there are major regional differences, the world has succeeded in feeding an increasing population over the last fifty years. Plant breeding, benefiting from significant budgets and the rapid integration of new techniques, has played a role in achieving this result.

Agriculture still faces a challenge. In addition there are now additional constraints due to the scarcity of new farmland, respect for the environment and the better integration of agriculture into the agro-food chain. Plant breeding is necessary in order to solve the problems raised. It should use the new techniques available, in particular plant transformation if the needs of a continually growing world population are to be met. The agro-food applications of the biotechnologies as well as their repercussions on human health will form a major part of economic and social development for the coming decade.

The test results on transgenic plants over the last twelve years demonstrate that, in general, the risk is extremely low compared with the benefits. To date, this has led to products being sold in large quantities throughout the world except in Europe, although we can expect this to change in the near future in the light of 1997 and 1998 decisions.

References

1. Jonson SP. *World Population Turning the Tide*. London : Graham & Trotman, 1994.
2. Klatzmann J. Alimentation, dans *Le Monde au Présent. Encyclopedia Universalis France*, 1994 : 25-34.
3. Wackemann G. Les stratégies agro-alimentaires. *Encyclopedia Universalis France*, 1989 : 125-31.
4. Feillet P, Rajnchapel-Messaï J. Agriculture, agro-industrie : un partenariat à maîtriser. *Biofutur* 1992 : 23-3.
5. *Biotechnologie, agriculture et alimentation*. Paris : OCDE, 1992.
6. Le Buanec B. Globalization of the Seed Industry. *Seed Sci & Technol* 1996 ; 24 : 409-17.
7. Desprez F, Devaux P. Les conséquences des biotechnologies sur le financement de la recherche. *Les biotechnologies du laboratoire aux champs : quels enjeux pour les semences ?* Compte rendu du colloque CENECA. Paris : Cultivar, 1989.
8. Joly P, Ducos C. *Les artifices du vivant*. Paris : INRA, 1993.
9. Cinquantenaire Limagrain, 1942-1992. Clermont-Ferrand : Imprimerie Reix 1992, et rapports annuels.
10. Craney J. Personnal communication, 1996.
11. Colomer JF. Cinquante ans d'aide au Tiers-Monde, la nouvelle approche des agriculteurs français. *La France Agricole* 1995 ; 2612 : 10-2.
12. Cauderon A. Espèces, variétés et semences dans l'évolution de la production végétale en France (1789-1989). In : *Deux siècles de progrès pour l'agriculture et l'alimentation*. Paris : Lavoisier, Académie d'Agriculture de France, 1990.
13. Persley GJ. Beyond Mendel's Garden; biotechnology in the service of world agriculture. CAB International, 1990.
14. Anonyme. Strengthening CGIAR. Private sector partnerships in biotechnology: a private Sector Commitee Perspective on Compelling Issues. 1997 : 22.
15. James C. Global Status of transgenic Crops in 1997. ISAA briefs n° 5 ISAA : Ithaca NY, 1997 : 31.
16. Les techniques de la transgénèse en agriculture. Académie des Sciences et CADAS, rapport commun n° 2. Paris : Lavoisier, 1993.
17. Knauf VC. Transgenic approaches for obtaining new products from plants, current opinion. *Biotechnology* 1995 ; 6 : 165-70.

2

A Historical Account of the French Biomolecular Engineering Commission

Sophie Béranger*, Hervé Reverbori, Eric Schoonejans*****

Summary

In 1986, France created the Biomolecular Engineering Commission in order to assess the safety of the deliberate release of genetically modified organisms.
The Commission includes scientific experts and is open to members of the public. It has acquired a great deal of experience, having examined 450 applications, representing over 3,000 release sites.
Most of these applications deal with plants.
It has also examined 13 applications for marketing approval, and this has led the Commission to acknowledge the need for a monitoring system for these cases in particular.

* Ex-Head of the Unit for Regulatory Affairs, Ministry of Agriculture, Fisheries and Food, Paris, France.
** Head of the Unit for Regulatory Affairs, Ministry of Agriculture and Fisheries, Paris, France.
*** Unit for Regulatory Affairs, Ministry of Agriculture and Fisheries, Paris, France.

From 1975, on both sides of the Atlantic, in France and in the United States, scientists who were beginning to imagine the possible applications of their discoveries in genetic engineering have met in order to assume a responsible attitude to the potential risks emerging from this technology. In the United States, the Azilomar conference set up an advisory committee to study and monitor the risks involved in the development and use of recombinant DNA: the Recombinant DNA Advisory Committee (RAC).

In France, the first deliberate release of Genetically Modified Organisms (GMOs) into the environment was accompanied by the establishment of an *ad hoc* scientific assessment commission. In 1986, the Ministry of Agriculture passed a decree establishing the Biomolecular Engineering Commission (in French, Commission du Génie Biomoléculaire: CGB).

The Law of July 13th 1992 concerning the use and release of GMOs confirmed the value of such a commission, named the "Commission for the study of the deliberate release of products derived from biomolecular engineering". This decree, n° 93-235 of February 23rd 1993, spelled out the tasks and composition of this commission.

The current members were named jointly by the Ministers of Agriculture and the Environment on 10 May 1993.

How does it Operate ?

Right from the start, the CGB associated scientific experts and members of the public. The legislation (Law of July 13th 1992) confirmed its composition. The CGB consists of 18 members including 11 scientists named for their competence in fields related to biomolecular engineering, one representative from an environmental defence association, one representative from industries using GMOs, one representative of the employees of these industries, a qualified person competent in legal matters and a member of Parliament in charge of scientific issues (in French: Office Parlementaire des Choix Techniques et Scientifiques). The CGB assesses the risks involved in the deliberate release and marketing of products that fully or in part consist of GMOs, and determines the conditions for use and their presentation. The secretariat is provided by the Ministers of Agriculture and the Environment.

The CGB must be consulted by the authorities about the deliberate release of any GMO. The President names two in-house examiners for each application. In addition, an outside expert is chosen by the applicant from a list of three scientists recommended by the President for any application concerning GMOs that have not yet been examined by the Commission. During the examination of the application, the applicant is asked to answer any questions that the members may have, and can make comments or suggestions. After a debate, a decision is taken, generally by consensus, exceptionally by vote. In addition, anyone can ask the CGB about any questions in its realm of competence. The CGB may provide decisions of general public interest. The annual report is sent to the Parliament.

A Historical Account of the Applications Examined

Research and Development

Since 1986, the CGB has examined 450 applications representing over 3,000 release sites. The distribution is indicated in *Table I* and *Figure I*. Each application may involve several GMOs, several release sites and/or several years.

Table I. Distribution of the applications examined.

	87	**88**	**89**	**90**	**91**	**92**	**93**	**94**	**95**	**96**	**97**	**Total**
Plants	5	8	13	24	31	27	40	51	72	115	123	509
Gene therapy	0	0	0	0	0	0	4	9	7	4	3	27
Recombinant vaccines	2	4	1	1	2	2	1	2	2	2	1	20
Recombinant products and microorganisms	2	5	4	1	3	3	3	6	1	4	4	36
Total	**9**	**17**	**18**	**26**	**36**	**32**	**48**	**68**	**82**	**125**	**131**	**592**

* On November 1998.

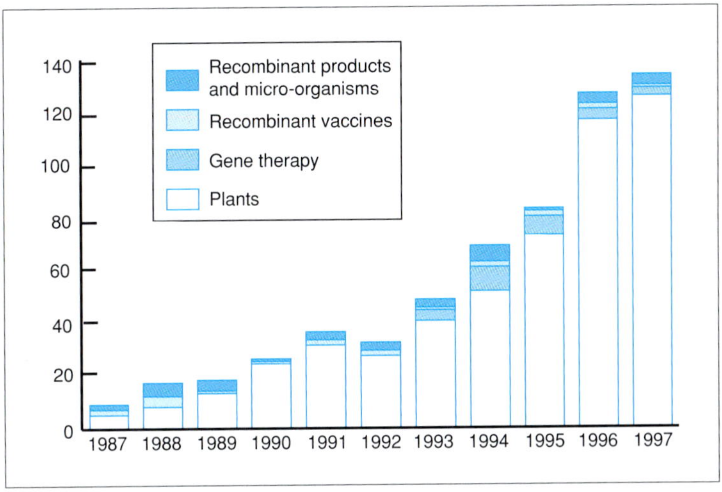

Figure 1. Distribution of the applications examined (on April 1998).

Since 1987, the number of applications has increased by more than 30% a year.

Most of the applications (84%) concern plants. However, applications for gene therapy have been increasing since 1993.

Fourteen species of plants of agricultural interest have been released since 1987 as indicated in *Table II*.

Table II. Release of genetically engineered crop plants since 1987.

	87	**88**	**89**	**90**	**91**	**92**	**93**	**94**	**95**	**96**	**97**	**Total**
Rape	0	1	4	8	7	7	12	16	23	29	13	120
Maize	0	0	0	0	5	6	11	15	21	37	23	118
Tobacco	3	6	6	8	5	4	5	4	8	12	5	66
Beet	1	0	1	4	7	5	7	7	9	11	7	59
Potato	1	0	0	1	3	1	2	4	1	1	0	14
Melon	0	0	0	1	1	3	1	2	0	3	0	11
Tomato	0	1	0	1	1	0	1	3	2	0	1	10
Poplar	0	0	1	1	1	1	1	0	2	2	1	10
Letucce	0	0	1	0	1	0	0	0	1	3	0	6
Sunflower	0	0	0	0	0	0	0	0	2	1	5	8
Chicory	0	0	0	0	0	0	0	0	1	1	0	2
Grape	0	0	0	0	0	0	0	0	1	1	0	2
Soybean	0	0	0	0	0	0	0	0	1	1	1	3
Gourd	0	0	0	0	0	0	0	0	0	1	0	1
Total	**5**	**8**	**13**	**24**	**31**	**27**	**40**	**51**	**72**	**102**	**51**	**424**

* Figures for April 1998.

Maize and rape are the most often tested plants. Tobacco, one of the first plants tested, is still relatively important. Maize, rape, tobacco and beet accounted for 84% of the field tests between 1987 and 1996 (*Figure 2*).

The characters introduced mainly involve tolerance to herbicides and resistance to diseases (insects, fungi and viruses).

Four types of characters account for 82% of the releases (*Figure 3*). No significant change in this balance has been observed over the last ten years.

Marketing Approval

Since 1993, the CGB has been asked to assess the applications to market genetically modified plants and veterinary vaccines. It has approved two

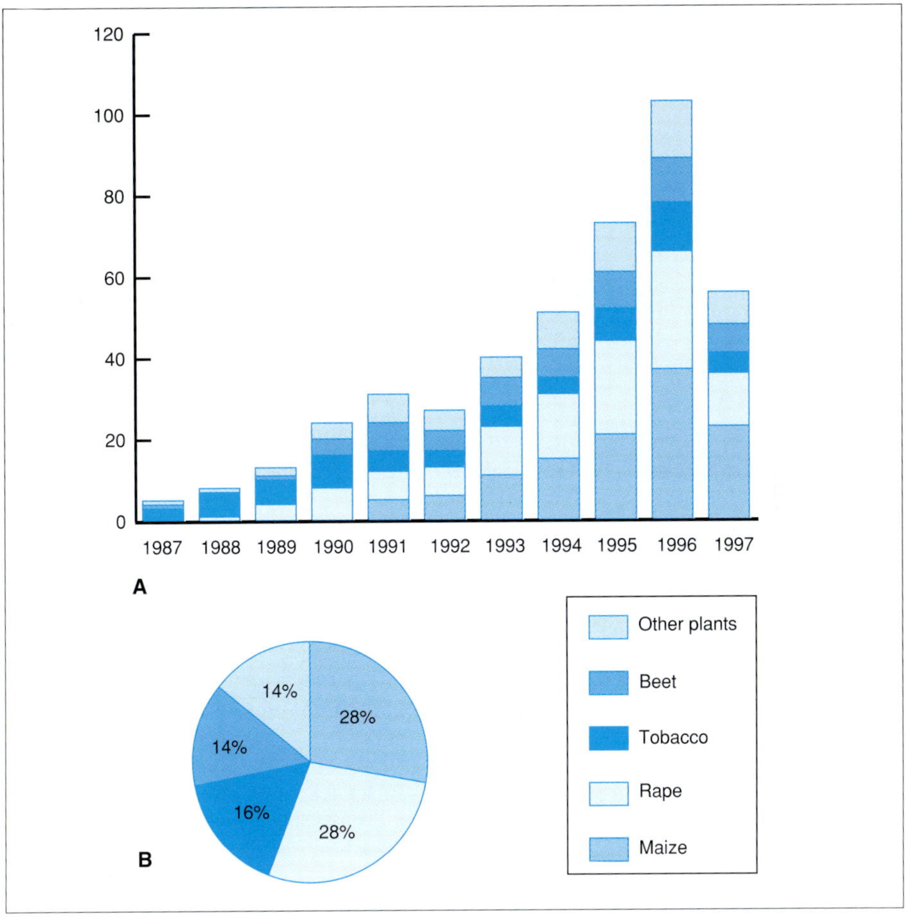

Figure 2. Distribution of the deliberate release of genetically modified plants per species characters introduced between 1987 and 1997.

applications for veterinary vaccines (vaccine against Aujeszky's disease, rabies vaccine) and thirteen applications for plants. The French authorities decided to provide EEC notification for nine of these applications (see below).

The Community authorities only approved six applications (two vaccines and four plants for limited use) (*Table III*).

The CGB is currently examining applications for six new plants.

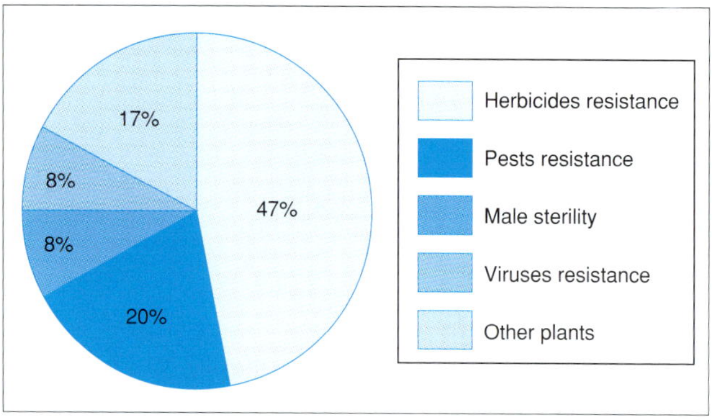

Figure 3. Distribution of the characters introduced between 1987 and 1997.

Table III. Applications for Genetically Modified Organisms (European Union).

Product	Use	Company	Notifying Country	State[1]
RABORAL	Vaccine	RHONE MERIEUX	Belgium	MC
NOBI PORVAC	Vaccine	INTERVET	Germany	MC
Herbicide-resistant tobacco	Limited	SEITA	France	MC
Rape (male sterility)	Limited	PGS	UK	MC
Insect-resistant maize	No restriction	NOVARTIS	France	MC
Chicory (male sterility)	Limited	BEJO	Netherlands	MC
Herbicide-resistant soybean	Import	MONSANTO	UK	MC
Rape (male sterility)	No restriction	PGS	France	CCE
Rape (male sterility)	No restriction	PGS	France	CCE
Insect-resistant maize	No restriction	MONSANTO	France	CCE
Herbicide-resistant maize	No restriction	AGREVO	France	CCE
Insect-resistant maize	No restriction	PIONEER	France	CCE
Herbicide-resistant rape	Import	AGREVO	UK	MC
Herbicide- and insect-resistant maize	Import	NORTHRUP KING	UK	MC
Blue flower carnation	Cut flowers	FLORIGENE	Netherlands	MC
Test kit	Milk specific	VALIO	Finland	MC
Chicory (male sterility)	Limited	BEJO ZADEN	Netherland	CCE
Herbicide-resistant rape	No restriction	AGREVO	Germany	CCE
Male Sterile Herbicide-resistant rape	No restriction	PGS AGREVO	Belgium	CCE
Herbicide resistant fodder beet	No restriction	DLF Trifolium	Denmark	CCE
Amylose free potatoes	Limited	AVEBE	Netherland	CCE
Low PG activity tomatoes	No restriction	ZENECA	Spain	CCE
Herbicide resistant cotton	No restriction	MONSANTO	Spain	CCE
Insect resistant cotton	No restriction	MONSANTO	Spain	CCE

1. MC: MC published in the Official Gazette of the European Community.
CCE: EEC approval still pending.

Other Activities

The experience acquired by the CGB during the examination of the applications led it to conclude the need to set up a plant monitoring system. The commercialization of these plants requires special attention as to the possible consequences on the land involved, for example, possible changes in farming practices or the effectiveness of phytosanitary products loss that are widely used and of major value to farmers.

In order to obtain data on the factors which will need to be monitored concerning the characters introduced in the commercial plants (herbicide-tolerance, European corn borer-tolerance, etc.), a study grouping the different technical centres involved (CETIOM, ITB, AGPM)*, INRA, industry and government officials has been undertaken. This large-scale study will be carried out on three sites under normal agricultural conditions and will last several years.

The need to set up a monitoring system following the commercialization of GMOs became clear during the meetings initiated by the CGB. These meetings involved both the concerned scientists and the industry representatives.

In order to try to answer the scientific questions that remain unsolved, the CGB launched a campaign to invite research projects, especially concerning the assessments of the risks involved in the release of transgenic plants.

Hence, whilst keeping up with the rapid progress of biotechnology and the problems related to the release and commercialization of different GMOs, the CGB has been able to develop, together with the scientific and industrial communities, a protocol to assess the safety of these activities. The large number of applications examined and its historical importance in the assessment of risks have given the CGB a unique breadth of experience recognized by the international community.

* CETIOM: Centre Technique Inter-Professionnel des Oléagineux Métropolitains.
ITB: Institut Technique de la Betterave.
AGPM: Association Générale des Producteurs de Maïs.

3

Analysis by the French Biomolecular Engineering Commission of the Potential Risks Associated with the Field Cultivation of Transgenic Plants

Axel Kahn*

Summary

The French Biomolecular Engineering Commission does not consider that genetic engineering used to improve plant varieties is by itself dangerous. Therefore, the potential risk involved in the cultivation of such transgenic plants may only be assessed on a case-by-case basis. The risk assessment procedure takes into account:
- the nature of the transgene,
- the genetic construct,
- the plant species and variety,
- the crop conditions and the ecosystem,
- the food or industrial use.
According to these different elements, the Commission successively assesses:
- the risks of toxicity,
- the food or feed risks for animals and humans,
- the risks of allergies,
- the ecological risks,
- the economic risks.

* Ex-President of the Biomolecular Engineering Commission, Paris, France.

Based on the potential advantage of a transgenic plant over non-modified plants in their natural or agricultural ecosystem, the transgenic plant is classified from 1 (no selective advantage) to 3 (high selective advantage).

Since the Biomolecular Engineering Commission (in French, Commission du Génie Biomoléculaire: CGB) does not consider that gene transfer in plants is dangerous in itself, the potential risk associated with the cultivation of transgenic plants can only be assessed on a case-by-case basis involving:
- the nature of the transgene,
- the genetic construct,
- the plant species and variety,
- the crop conditions and the ecosystem,
- the food or industrial use.

Different types of risks are assessed on the basis of these elements.

Risks of Toxicity

They may be directly related to:

- the nature of the product whose synthesis is controlled by the transgene. This toxicity is often well documented in the scientific and technical literature or may be determined using standard toxicology techniques;

- the change in the metabolism and thereby the composition of the plant resulting from the biological effects of the gene transfer. Depending on the transgene used (for example, a gene coding for an enzyme that may interfere with certain metabolic pathways or attack certain plants components), the composition of the transgenic plant may differ from untransformed plants and thereby call for examinations to determine the consequences;

- the modified metabolism of chemical substances, mainly herbicides, that the transgenic plant tolerates. Genes for herbicide resistance may make the plant insensitive to the herbicide or provoke its rapid degradation. In the first case, the persistence of the herbicide in a plant on which this herbicide has never been used before may lead to the appearance of novel metabolites. In the second case, the enzymatic degradation of the herbicide will also give rise to new metabolites that may undergo specific degradation in their turn. It is necessary to measure these new compounds and assess their toxicity quantitatively and qualitatively;

- the high level of pathogen infestation of a genetically tolerant plant that, theoretically, could be accompanied by the accumulation of potentially toxic substances.

Food or Feed Risks for Humans or Animals

Other than the toxic risks mentioned in the paragraph above, this problem can only occur when the composition of the transgenic plant is not substantially identical to that of non-transgenic varieties currently used. In addition, it is necessary to check that the transgene does not have an effect on the level of expression of anti-nutritive plant substances known to exist in the variety under consideration. This is described in the chapter devoted to food risks.

Risks of Allergies

These risks are more difficult to determine except in simple cases where the transgene comes from a species that is known to involve a risk of allergic reactions, or even codes for an already identified allergen. If not, the assessment may be based on the structural similarities between the product of the transgene and known allergens and on the residual levels of the protein coded by the transgene in the product for consumption. There now exist data banks for potentially allergenic peptide motifs that facilitate this study. Often however, only "biomonitoring" will eventually resolve this question.

Ecological Risks

The proliferation of a transgenic plant in the ecosystem or of an interfertile species that acquires highly advantageous genes is a major question for discussion. This will be fully dealt with in other chapters in this book.

The effect on the equilibrium of domestic and wild insect populations is also discussed in the chapter devoted to the introduction of genes coding for *Bacillus thuringiensis* toxins. It is difficult to carry out this study in experimental conditions beyond the search for an acute toxic effect.

The selection of more virulent pathogens or pathogens with an altered host range by the use of genetically resistant plants is also a problem that will be dealt with elsewhere in this book.

The horizontal transfer of DNA, in particular to soil bacteria does not seem to involve a major risk. However, the genetic constructs should not increase the theoretical probability of the phenomenon.

Based on these elements, transgenic plants may be classified into three different categories [1]:

- **category 1** corresponds to transgenes that are not likely to give a selective advantage to the plants in which they are transferred: male sterility factors, withering inhibitors, enrichment in a specific amino acid, etc. The first two of these characteristics may even be selectively disadvantageous;

- **category 2** corresponds to transgenes that may provide a slight selective advantage in specific growing conditions. Sub-category 2a includes characteristics that do not provide any specific advantage outside of an applied selection pressure in an agricultural context (herbicide resistance genes). Sub-category 2b includes characteristics with an advantage under almost all conditions, for example, genes for resistance to bacteria, viruses, yeast, insects, different types of stress, etc.

- **category 3** includes all of the characteristics that are likely to provide a high selective advantage to plants expressing them: genes increasing strength, fertility, accelerated flowering, etc. The influence of this classification on the assessment of the ecological risk will depend on the sexual characteristics of the variety (allogamous or autogamous), the pollination distances involved, the risk of seed dissemination, seed resistance and dormancy, the existence of interfertile wild species, the invasiveness of the varieties from which the transgenic plants are derived, etc.

The "Economic" Risks

This is a difficult chapter. The assessment of these risks is at the outer limit of the mandate of the CGB. However, with ten years of experience, the CGB has identified possible economic risks which should be avoided in the development of plant biotechnologies. Commercial failure resulting from the fears and doubts of the population concerning transgenic plants and their by-products is an economic risk that is best identified by industrialists. In order to avoid this outcome, the questions and concerns of the consumers have to be answered as honestly and clearly as possible.

The CGB and other scientists and industrialists involved have discussed over the last few years additional potential economic problems related to the large-scale cultivation of transgenic plants. One of them involves the risk of a loss of efficiency of total herbicides (often presented as "acceptable" herbicides concerning their effect on the environment) which will result from poorly controlled dissemination of transgenes conferring resistance to these herbicides. A similar problem may arise if the use of plants producing insect toxins favours the emergence of resistant insect populations.

It is important to identify such undesirable economic effects and define the best way to avoid them. The future of plant biotechnology depends on its economic viability. In addition, this aspect is one of the legitimate concerns of the farmers employing transgenic plants. Obviously, the CGB has to try to answer these questions. This is a realm where clear-cut predictions are not possible since the technical, economic and social factors involved are complex and intangible and the view of such risks may be subjective. The position of the CGB is to alert all of the actors involved in the production and use of transgenic varieties about the possibility of such risks and favour the discussion of the best ways to avoid them.

Reference

1. Ahl Goy P, Duesing JH. Assessing the environmental impact of gene transfer to wild relatives. *Biotechnology* 1996 ; 39-40.

4

Transgenic Plants and Food Safety

Gérard Pascal*

Summary

The evaluation of the safety (both nutritional and toxicological aspects) of foods and ingredients derived from transgenic plants is based on the concept of substantial equivalence. The idea behind this concept is to compare these novel foods with comparable existing foods, which are traditionally consumed without any undesirable effects.

Further safety assessment is no longer required when substantial equivalence is proven, based on the analytical data (typical levels of nutrients, anti-nutritional or toxic compounds) and on a precise knowledge of the nature of the introduced genetic constructs.

When the presence of the products of the genes of interest and introduced markers is the only difference between the novel food and its reference, the safety of these products must be specifically assessed. Particular attention must be paid to the risks of allergies due to the presence of new proteins as well as to the risks of secondary effects due to the products of the inserted genes, which may modify the plant's metabolism. In most cases, this assessment may be carried out by chemical analysis, in vitro assays, and sometimes acute or short-term toxicity tests when necessary.

Finally, when substantial equivalence cannot be demonstrated, sequential evaluation is necessary. This may range from a simple analytical study to a full toxicology assessment.

* Member of the CGB, Senior Scientist, Head of the Centre National d'Études et de Recommandations sur la Nutrition et l'Alimentation (CNERNA-CNRS), Paris, France.

The difficulties in using classic toxicology studies for foods are emphasized, as is the need for careful consideration of the experimental conditions to be used when these studies are required.

In 1983, the Science and Technology Committee of the OECD (Organization of Economic Co-operation and Development) set up a National Experts Group (NEG) to assess the safety of biotechnology. The NEG decided that priority should be given to work on food safety, in particular the safety of novel foods and food ingredients produced by biotechnology. A specialized working group was created. It drew up the general principles published by the OECD in 1993 under the title: *Safety Evaluation of Foods Derived by Modern Biotechnology: Concepts and Principles* [1]. This was the first document to use the concept of substantial equivalence. Since then, this concept has been used on an international basis to draw up strategies for assessing the safety of novel foods. This concept is based on the idea that comparison with existing organisms used as food or food sources may be used to assess the wholesomeness for human consumption of a food or ingredient that has been modified or is new. Wholesomeness includes both nutritional and toxicological aspects.

The idea of comparing a novel food with a conventional safe food traditionally consumed without any undesirable effects was raised during a FAO/WHO meeting in Geneva in November 1990 concerning the "Strategies to Assess the Wholesomeness of Foods Produced by Biotechnology" [2,] although the concept of substantial equivalence was not clearly defined.

In 1994, the OECD and the WHO organized two meetings to better determine the possible uses of this concept. The goal of the first meeting (Oxford, 12-15 September) was fairly ambitious since it aimed at laying the foundations for the safety assessment of novel foods and the new technologies in general [3]. The second meeting (Copenhagen, 31 October-4 November) was much more focused. It was devoted to the *Application of the principles of substantial equivalence to the safety evaluation of foods or food components from plants derived by modern biotechnology* [4].

This contribution is based on the discussions initially undertaken by these international agencies (continuing during the following months) as well as on the experience acquired during the examination of the files in France and in different member states of the European Union or what is known of the cases examined in the United-States. It also takes into account the first results provided by certain European research programs (in particular FLAIR). The position taken by the Biomolecular Engineering Commission (in French: CGB) will be outlined.

The characteristic principles of the concept of substantial equivalence (SE) are:
- SE takes several key compounds in the composition of a product into account (natural toxins, typical nutrients, anti-nutritional factors) compared to those of the traditional reference product;
- SE is a dynamic concept since a novel food accepted as equivalent may in turn be used as the basis for evaluation (reference);
- the assessment of SE may facilitate the identification of particular aspects of the novel food requiring additional study;
- to determine the SE of a genetically modified organism (GMO), it may be compared with the parent organism alone or with different varieties of the same species;
- agronomic and phenotypic characters are included in the assessment of SE.

Several major parameters to be analysed for establishing SE have been identified:
- the molecular characterization:
    - inserted DNA (source, function, sequence);
    - product and sites of gene expression;
- the agronomic features usually assessed for inclusion in the catalogue;
- the chemical characterization (key nutrients, natural toxins, etc.);
- the identification and characterization of an appropriate reference;
    - the analysis of the key components;
    - the characterization of the reference taking into account the natural variability due to genetic factors, environmental factors, post-harvest treatments, processing, the potential uses of the new product and its role in the food;
    - the standardization and validation of the analytic methods;
    - reference to data bases.

Assessment of Substantial Equivalence

There are three possible situations.

The equivalence can be demonstrated

In this case, no difference can be demonstrated between the food products derived from the transgenic plant and those from a conventional counterpart. This may be the situation when the desirable genes or markers are not expressed in the edible parts of the plant. In these conditions, the CGB considers that no other demonstration of the wholesomeness is required. The transgenic plant or its products are considered as safe food, just like the conventional counterpart or its products.

The equivalence is demonstrated, except for the presence of products from the introduced genes of interest or marker genes

In this case, it is necessary to demonstrate the safety of these products or metabolites resulting from their action.

A major concern is the potential for human allergies to these products, most often proteins. The OECD consultations have examined this aspect. An initial screen has been recommended by the OECD that takes into account the current lack of tests to directly demonstrate such a potential. The following criteria have been selected:
- the molecular weight of allergen proteins is generally between 10 and 70 kDa;
- the amino acid sequence of these potential allergens is similar to that of proteins known to be allergens. The specialists consider that it is necessary to have at least one stretch of sequence identity involving at least eight adjacent amino acids to be significant from the immunological point of view;
- they resist heat denaturation and digestion;
- their level of glycosylation and their abundance are important factors in the assessment.

All these criteria may be assessed using biochemical analyses and *in vitro* tests carried out in a reconstituted gastric and intestinal environment. An *in vivo* acute toxicity test may be added. It is difficult to predict the allergenic potential of proteins. Post-marketing monitoring of transgenic plants and/or their products is often recommended as an efficient way to monitor foods as long as these products can be distinguished from similar unmodified products. Their "traceability" is thus involved. Food monitoring is not only of interest in the case of products derived from GMOs, but in general for all novel or exotic food products that may contain poorly known allergens that are not commonly encountered due to our usual eating habits.

When the proteins produced by the plant have an enzymatic activity that, for example, enables the metabolism of a pesticide, it is obvious that the toxic potential of the metabolites should be studied. This assessment is systematic for the authorization of pesticide treatments granted for specific uses (plant/product pair) by specialized commissions (in France by the Commission d'Étude de la Toxicité and the Comité d'Homologation des Produits Antiparasitaires à Usage Agricole et des Produits Assimilés).

The assessment of the secondary effects of gene insertion is another difficult question. Certain genes transferred to plants code for enzymes that catalyse particular steps in biochemical processes. The expression of the genes may lead to the depletion of an enzyme substrate or the accumulation of the product of the reaction or an increase in the flow of metabolites in the subsequent steps.

These effects are to a large extent predictable and must be assessed on the basis of, for example, the knowledge of the specificity of the enzyme for the substrates that are naturally found in the plant. They are not a new health risk since identical situations may arise following mutagenesis when using traditional methods of selection.

The theoretical risk of mutagenesis by insertion should be considered. The insertion of a gene may modify or interrupt the usual expression of existing genes in the plant. There may be inactivation following insertion into coding regions or activation after insertion into regulatory regions. However, since most of the DNA in plant genomes is made of non-coding or repetitive elements, mutagenesis by insertion will be rare. Hence DNA insertion will probably not generally have a phenotypic effect and when it does, this will usually be the result of gene inactivation. As regards wholesomeness, the main potential danger of these mutageneses is the activation of silent or barely expressed genes that may result in the biosynthesis of toxic secondary metabolites.

Although this question is easily dealt with by chemical analysis in the case of toxins or anti-nutritional compounds that are naturally found in certain species (solanin in potato, glucosinolates in rape, tomatin in tomato, etc.), it is more difficult to determine the theoretical risk involved in insertional mutagenesis due to the production of new toxins by activation of normally inactive metabolic pathways. Chemical analysis only finds what is searched for *a priori*. There are two possible solutions to this problem:
- comparison of the "analytic profiles" of extracts specifically made to obtain compounds of known toxin families naturally found in plants;
- animal testing to reveal the toxic effects of the transgenic plant or its products.

The second solution has been used in several cases during the last few years. The results were often disappointing, as in the case of the FLAVR-SAVR tomato from Calgene. Rats were force-fed powdered freeze-dried transgenic and non-transgenic tomatoes. This technique is tricky and led to accidents that sometimes arose more often in the "transgenic tomato" group. It was long and difficult to demonstrate the randomness of these events [5, 6]. However, nothing abnormal was observed with another transgenic tomato expressing the product of a *Bt* gene when the powdered tomato was added to rats' food [7].

Another example involves transgenic rape tested on rats. Rats are particularly sensitive to the toxic or anti-nutritional compounds in rape, especially glucosinolates. It is difficult to obtain control rape with the same composition as transgenic rape when the latter is obtained from a single seed while the parental variety consists of individuals with different properties. This is true in North America where the varieties consist of populations and not, as most often in Europe (especially in France), of pure or practically pure registered strains. In addition, the time, place and agricultural conditions under which the crop is grown will influence the composition of the seeds obtained. Thus it is always possible to observe differences between groups of rats although it is not possible to attribute these differences to the risk due to genetic engineering.

Ideally, comparisons should be conducted between a genetically modified organism and an isogenic conventional counterpart (isogenic apart from the transgenes) grown under the same conditions.

Recent international meetings have noted the disadvantages of animal tests on complex foods. These tests are not sufficiently sensitive to detect unintentional effects of genetic modifications. Only a reasonable dose of food can be added to the diet of test animals in order to avoid nutritional imbalance, meaning that it is not possible to use overdoses to evaluate a safety margin as is done when studying food additives or contaminants. No agencies recommend the systematic use of classical rodent toxicology tests.

However, animal feed studies may provide useful information concerning the wholesomeness of transgenic plants. These studies may be carried out in species that usually consume the test product under normal feeding conditions. The usual parameters may be monitored, especially the rate of growth, feed consumption and the index of consumption. These parameters are highly sensitive in certain species during the period of rapid growth when all of the animals' potential is mobilized. It is also possible to measure the weight of the major organs (liver, kidneys, testes if sexual maturity is reached, etc.) and store tissue and organs for a histology study if disturbing indications are obtained after the examination of the other parameters.

These conclusions are not only derived from the experience acquired in the assessment of transgenic plants but also during the long and numerous studies carried out to determine the effects of food irradiation [8].

The first solution, involving thorough chemical analysis, should be explored more extensively. Such recommendations have been made, especially by Noteborn *et al.* [9]. They recommend the "metabolic imprint" concept to detect any secondary effects due to the genetic modification. This approach involves the use of modern techniques of physicochemical analysis (all types of spectrometry and chromatography) and knowledge of the families of toxic substances naturally found in the plants. It is highly improbable that these secondary effects involve the production of a substance that has never been observed in the plant world before. Differences noted in the imprint of certain extracts between transgenic plants and control plants may then give rise to more extensive studies to identify the offending substance. If, exceptionally, the substance is not already known, it is possible to assess its possible toxicity using classical toxicology tests.

It will be necessary to test this methodology, as it is still at the proposal stage. If successful, it would satisfy any theoretical objections without resorting to animal tests that are of uncertain usefulness in these cases. This approach is similar to that sometimes used to test the toxicity of food flavourings from natural extracts, or the analysis of the fatal accidents arising in the United-States after consumption of tryptophan from batches presenting an unusual chromatography peak.

The substantial equivalence between a novel food and a similar traditional food cannot be determined

This is the case when the gene(s) introduced code for a character that modifies the plant as regards its use as food, for example the production of a new oil or a new carbohydrate. A case-by-case approach must be used, based on the properties of the novel food. Animal toxicology tests may be required, although not systematically, since the lack of equivalence does not necessarily mean the food is unsafe.

Conclusion

Substantial equivalence is very useful in assessing the wholesomeness of transgenic plants and their products. International consensus has not yet been obtained as regards its interpretation and use. The conclusion following the demonstration of substantial equivalence is that the transgenic plant has the same degree of wholesomeness as a traditional reference food that is considered to be risk-free on the basis of a long experience of consumption. This is not a traditional toxicology assessment since irradiated foods were the first foods to be assessed in this way. Ordinary foods have never been assessed in this manner.

Based on current experience, the BEC feels that:

- substantial equivalence may be assessed by means of an analysis of the characteristic nutrients in the plant or the derived products consumed by humans or animals, as well as the anti-nutritional or toxic compounds that are known to be found in the species under consideration;
- when equivalence is demonstrated, apart from the products of desirable genes or introduced markers, the safety of these products should be specifically assessed. This assessment may involve *in vitro* tests as well as *in vivo* acute toxicity tests. For these tests, it may be necessary to over-express the product of the transgene in a micro-organism when it is not possible to obtain sufficient quantities from the plant. Obviously, the identity of the products has to be demonstrated;
- the demonstration of the wholesomeness of transgenic plants should not systematically rely on classical toxicity tests on rodents as they are often not suitable. Two different situations may occur, according to the manner by which the secondary effects of gene insertion should be assessed.

In the first situation, the product of transgene expression is well known, its role is highly specific and does not seem likely to modify the metabolism of the plant, (for example, the expression of *Bacillus thuringiensis* toxin Cry I A). In these conditions, toxicological testing is not required. Tests can be used to demonstrate the efficiency and safety of the plant products in animal nutrition. This provides sufficient additional information to assess the substantial equivalence.

In the second situation, the product of the gene expression is an enzyme whose action may modify plant metabolism (modification of the production of biopesticides, hormones, toxic components). Toxicological testing is then useful along with in-depth analytical tests. The conditions for such testing should be carefully defined so that the results can be interpreted.

References

1. Organisation de Coopération et de Développement Économiques. *Évaluation de la sécurité des denrées alimentaires issues de la biotechnologie moderne : concepts et principes*. OCDE, 1993 : 87 p.
2. Organisation Mondiale de la Santé. *Stratégies d'évaluation de la salubrité des aliments produits par biotechnologie*. Rapport d'une consultation conjointe FAO/OMS, OMS. Genève, 1993 : 66 p.
3. Organisation de Coopération et de Développement Économiques. *Food Safety Evaluation*. OECD Documents. OCDE, 1996 : 180 p.
4. World Health Organization. *Application of the principles of substantial equivalence to the safety evaluation of foods or food components from plants derived by modern technology*. Report of a Workshop. WHO, 1995 : 78 p.
5. Redenbaugh K, Hiatt W, Martineau B, Emlay D. Regulatory assessment of the FLAVR SAVR tomato. *Trends Food Sci Technol* 1994 ; 5 : 105-10.
6. Hattan D. Evaluation of Toxicological Studies on FLAVR SAVR Tomato. In : *Food Safety Evaluation*. OECD Documents. OCDE, 1996 : 58-60.
7. Kuiper HA, Noteborn HPJM. Food Safety Assessment of Transgenic Insect-resistant *Bt* Tomatoes. In : *Food Safety Evaluation*. OECD Documents. OCDE, 1996 : 50-7.
8. Hattan D. Lessons Learned from the Toxicological Testing of Irradiated Foods. In : *Food Safety Evaluation*. OECD Documents. OCDE, 1996 : 11-21.
9. Noteborn HPJM, Bienenmann-Ploum ME, van der Berg JHJ, Alink GA, Zolla L, Kuiper HA. Food safety of transgenic tomatoes expressing the insecticidal crystal protein CRY IA (b) from *Bacillus thuringiensis* and the marker enzyme APH 3'. *II Med Fac Landbouww Univ Gent* 1993 ; 58/4b : 1851-8, and personal communication.

5

The Creation of Transgenic Plants

Francine Casse*

Summary

Genetically modified plants are of huge interest in basic research since it is the only way to study the different levels of regulation in gene expression. The knowledge obtained can be quickly used to introduce new traits of agricultural interest such as resistance against herbicides or pests, modified protein or lipid composition, male sterility, etc. However, the potential applications of plant transformation are not restricted to plant breeding. They are also of interest to industry, where plant cells or whole plants are thought of as novel factories, consuming solar energy and carbon dioxide and producing proteins of high added value. Successful plant transformation involves satisfying several requirements: penetration of foreign DNA into plant cells, integration into the genome, expression of the transgenes, and regeneration of whole plants from the genetically modified cells. The most common DNA transfer technique involves the use of the natural ability of the soil bacterium Agrobacterium *to transfer a defined DNA region into the plant nuclear genome. The vectors for transformation* via Agrobacterium *can be divided into two groups: cointegrative intermediate vectors and autonomously replicating binary vectors. Most important crops are monocots, which for a long time were considered to be insensitive to* Agrobacterium. *In consequence, various direct transfer techniques were developed, in particular biolistics, in which*

* Member of the CGB, Professor, Université Montpellier II, Montpellier, France.

a gun is used to project DNA-coated microparticles into the tissue to be transformed.
In addition to nuclear transformation, other techniques are being developed, such as organellar transformation, which will allow maternally inherited traits to be introduced, or viral vectors which will allow scientists to express new genes during the life of the plant.
The molecular characterization of transformants is achieved by Southern transfer, molecular probing and PCR amplification.
Future progress in transformation techniques should include the extension of their application to tropical plants, and the development of methods to eliminate the marker genes that are only useful in the primary selection procedure.

Background

The history of plant transformation began with the discoveries made by phytopathologists working on soil bacteria responsible for crown gall and "hairy root" syndrome, namely *Agrobacterium tumefaciens* and *A. rhizogenes* respectively. At the INRA in Versailles, Georges Morel and his team first demonstrated that the tumours or roots that these bacteria induce on sensitive plants synthesize special substances, called opines. Opines are strain-specific, and are growth substrates for the bacteria that produce them [1]. In 1974, Jeff Schell and Marc Van Montagu and their team in Belgium demonstrated that this transformation of plant cells is due to plasmids found in the virulent strains of *Agrobacterium* [2,3]. These plasmids are called Ti (Tumour-inducing) in *A. tumefaciens* and Ri (Root-inducing) in *A. rhizogenes*. In 1977, Mary Dell Chilton, along with Eugen Nester's team in the United-States, demonstrated that the transformation of plant cells by *Agrobacterium tumefaciens* results from the incorporation into the plant genome of a DNA fragment (called T-DNA for transferred DNA) derived from the Ti plasmid [4]. In 1982, M.D. Chilton, this time with an INRA team (Versailles) also showed that the roots induced by *A. rhizogenes* contain T-DNA from the Ri plasmid in their nuclear genome [5].

Since the discovery of the Ti and Ri plasmids, studies have shown that the DNA transfer mechanisms are identical in the two bacteria and that the pathogenic effect of these bacteria is due to the characteristics of the transferred genes that perturb the hormone metabolism of the transformed cells. The molecular mechanisms underlying these natural genetic transformation phenomena are becoming clear. They resemble bacterial conjugation but in this case the host is a plant cell [6]. The genes carried by the T-DNA are not expressed in *Agrobacterium* but only in the plant cell nucleus as they carry eukaryotic regulation signals. To obtain transgenic plants, the oncogenes are removed from the T-DNA, giving rise to so-called "disarmed" vectors. The now non-pathogenic plasmids retain the ability to participate in gene transfer (but in this case only the desired genes are transferred) from a bacteria towards the plant cell nucleus.

Parallel to this research, direct DNA transfer techniques using chemical or physical methods or electric shocks were developed, mainly with animal cells. Subsequently, these methods were tested on protoplasts, which are plant cells with their rigid cell wall removed. Thanks to these techniques, provided that the DNA reaches the nucleus, one can obtain the synthesis of the proteins encoded by the transferred genes, as long as the transgenes are able to be expressed in the host cell. The first transgenic monocots were obtained in this way.

The success of plant transformation depends on optimizing several parameters: the penetration of foreign DNA into the plant cells, the ability of the transgenes to be expressed, the integration into the host genome, the ability of the genetically modified cells to regenerate whole plants. All of these conditions were met for the first time in 1983. The team in Ghent obtained transgenic tobacco plants expressing a chimaeric gene making them resistant to an antibiotic, kanamycin [7,8].

The transfer of DNA into a cell, even if it reaches the nucleus, does not necessarily mean that it will be expressed. To lead to the synthesis of a protein, the gene sequence has to have signals that can be recognized by the host cell machinery and, first and foremost, promoter signals that are recognized by the RNA polymerase responsible for transcribing messenger RNA, and so-called enhancers or silencers, recognized by the nuclear proteins that regulate the rate of transcription. The messenger RNA then has to be recognized by the ribosomes, responsible for translating the genetic code into proteins. Finally, once translated, the proteins are targeted to the cell compartment where they function. All of these mechanisms and all of the signals involved differ between prokaryotes (organisms without a nucleus, *i.e.* bacteria) and eukaryotes (organisms that contain a true nucleus like animals and plants) or even often between animals and plants. However, almost any protein, whether animal, bacterial or viral may be perfectly well synthesized in a plant, provided that the nucleotide sequence encoding its amino acid sequence is surrounded by signals that will allow its expression (transcription, translation, targeting) in plant cells.

Using now classical molecular biology techniques, this simply requires the *in vitro* construction of a gene. Such recombinant genes are said to be chimaeric, since the sequence may be built from a mixture of animal, bacterial, viral or plant elements.

However, the transfer of genes able to express themselves in plant cells does not necessarily give rise to transgenic plants. It suffices for the DNA to reach the nucleus for it to be transcribed into RNA, and once this RNA is exported to the cytoplasm it will be translated into protein leading to transient expression of the introduced genes. However, stable expression of the newly introduced DNA will only occur if it becomes integrated into one of the chromosomes, such that the transgenes can be replicated along with the host DNA. In certain cases, for example when it is only necessary to find out if a new chimaeric gene construct is functional, it is possible to simply measure transient expression. This can, for example, be done using protoplasts into which the transgene has been transferred, even if these protoplasts are not able to divide. However, to obtain a transgenic plant with cells possessing the transgene, the transforming DNA must be integrated within the nuclear genome and thereby be transmitted to the daughter cells.

Occasionally, the DNA may become integrated into the genome and the transformation may be stable, but only an undifferentiated callus can be derived from the initially transformed cell. To obtain a transgenic plant, a whole individual also has been able to regenerate from the initially transformed cell. Therefore, last but not least, whatever the method of transformation used (*via Agrobacterium* or by direct transfer), the target of the DNA transfer should be a type of cell able to give rise to an entire organism.

Finally, since gene transfer is never 100% effective, it is necessary to select the genetically transformed regenerants from all the other potential regenerants. If the expression of the desired transgene is not sufficient for the early selection of the transgenic cells or individuals from those that are not, it is necessary to transfer a selectable marker gene along with the desired gene. The transformed individuals are then selected on the basis of the expression of the marker gene. The most commonly used selection markers are chimaeric genes conferring resistance to an antibiotic that is toxic for plant cells (kanamycin, hygromycin) or a herbicide [7].

The Goal of Transformation

Plant transformation is now a widely used tool, both in basic research where it is used to study the regulation of the expression of plant genes, or facilitate their cloning by tagging, and as an additional technique in plant breeding to complement classical genetic methods or for industrial purposes to produce plants containing substances of high added value.

Basic Research

Transformation is above all a powerful tool for genetic and physiological analysis [9]. Its use in this area is increasing.

Transgenic plants are used to study the molecular mechanisms that control tissue- or organ-specific expression (leaf, flower, tuber, seed) or induction of transcription under certain conditions (light, temperature). Chimaeric genes can be constructed with the coding sequence of a reporter gene, coding for an easily quantifiable stable protein, under control of the regulatory elements of the gene to be studied. These genes can be modified by *in vitro* mutagenesis before introducing them into plants where their expression will be monitored. It is therefore possible to create mutations in regulatory regions that could not have been isolated directly from a whole organism and study their effect by assaying the expression of the reporter gene [10]. Cis-acting transcription regulatory elements of an increasing number of genes are being identified and studied in detail.

Although it is still not possible to introduce mutations into a specific gene *in situ*, it may be inactivated by expressing antisense RNA in the plants that may inhibit it by base-pairing with it [11]. Co-suppression is another way of preventing the expression of a gene [12,13]. This phenomenon was unknown (and unsuspected) before the widespread use of plant transformation. In this case, the presence of an additional copy of a gene does not increase expression as would normally be expected but, on the contrary, suppresses both the resident gene and the transgene. This unexpected finding has been a rich source of results on the control of gene expression which several groups have been eager to exploit.

The integration of T-DNA into a nuclear gene may inactivate it or make the transgene depend on the regulatory signals of the target gene. Using appropriate vectors, this property is now mainly used for gene tagging [14-16].

Applications

Increasingly numerous examples of the transfer of genes of agricultural value illustrate the applications of genetic transformation. Plants have been developed with new characteristics after transfer of chimaeric genes with coding sequences derived from plants as well as animals or micro-organisms or even chemical synthesis. The most studied traits include resistance to herbicides, insects or phytopathological organisms such as viruses, bacteria or fungi.

Three strategies can allow a cell or organism to resist a toxic compound: over-expression of the target compound, expression of a mutated resistant target compound, and expression of an enzyme able to transform the toxic product into a non-toxic derivative. As will be seen in the chapter devoted to this subject, these three types of resistance are used to obtain plants that are resistant to different families of herbicides.

In 1987, Plant Genetic Systems [17], Agrevo [18] and Monsanto [19] independently published that they had obtained insect-resistant transgenic plants by the expression of a toxin normally synthesized by *Bacillus thuringiensis*. As will be described later on in this book, *Bacillus thuringiensis* is a source of insecticides that is far from being exhausted. However, insect-resistance may also be obtained by the expression of protease inhibitors, such as those synthesized in the seeds of certain leguminous plants [20] or in wounded plants [21], or α-amylase inhibitors [22].

Different strategies have been developed to make plants virus-tolerant. Most of them are based on the concept of parasite-derived resistance [23], and aim at perturbing the viral life-cycle by the presence of a viral RNA or protein in the wrong place or at the wrong time. In 1986, Roger Beachy, together with Monsanto, demonstrated that expression of the tobacco mosaic virus coat protein gene in transgenic tobacco delays the appearance of symptoms after infection by this virus [24]. Since then, the list of virus-tolerant plants due to the expression of the capsid protein has not stopped expanding. Resistance may also be obtained by the expression of other viral proteins, in mutated non-functional forms, such as the replicase [25,26] or movement protein [27]. Strategies not relying on protein synthesis are also being studied, such as the expression of antisense RNA that may block translation, sense or defective interfering RNA, likely to lure the replicase away from the viral RNA, or satellite RNA whose replication competes with the virus itself. These different strategies and their associated potential risks are discussed elsewhere in this book.

Different chimaeric genes have been built to obtain resistance to fungi and bacteria [28]. Resistance can be obtained by the synthesis in the transgenic plants of chitinase [29], glucanase [30], ribosome-inactivating protein [31], phytoalexin [32,33], lysozyme [34] or other antibiotic proteins that are normally produced by insects [35,36].

Using genetic engineering, it is also possible to modify the amino acid composition of the storage proteins in seeds and thereby improve their nutritive quality or technological potential. The Brazil nut gene coding for a methionine-rich storage protein was transferred to rape [37] and a gene coding for a phytase was used to improve the use of phosphorus in feed [38]. The starch content may be modified in diverse ways [39].

Transformation may be used to change the taste of products. It suffices to introduce a gene coding for a protein that, like thaumatin or monellin, has a sweet taste [40,41].

The introduction of genes coding for enzymes involved in lipid metabolism may change the length of the fatty acid chain or the position or number of double bonds, thereby improving the quality of oils used for food or other purposes [42-44]. The short-term goals may be to obtain rape and soybean where 80% of the fatty acids have a medium-length chain, oleic acid accounting for over 90%, with a modified erucic acid content and containing ricinoleic acid [42].

Transgenes modifying the expression of the enzymes intervening in the biosynthesis of flavonoids change the colour of the flowers [11, 45-47], although this is more often done to understand the control of pigment synthesis than for commercial reasons. Calgene was one of the first companies to develop transgenic tomatoes with altered ripening and texture [48]. Different strategies are being used to try and improve the preservation of fruit, such as the inhibition of polygalacturonase [49], pectin methylesterase [50], the two last enzymes in the biosynthesis of ethylene [51,52] and more recently, the over-expression of a cytokinin synthesis gene [53].

The production of hybrids may also benefit from transformation techniques.

Different strategies for creating male sterility are being studied. The male sterility/fertility restoration system developed by Plant Genetic Systems using expression of a specific ribonuclease strictly limited to tapetum cells [54], or a ribonuclease-inhibitor specific to this ribonuclease [55], is the most advanced.

We will not detail them here, but many studies have been developed to identify and then transfer to plants the genes involved in tolerance to environmental conditions limiting productivity such as drought, cold, oxidative stress, salt, etc.

Since the advent of plant transformation, plant breeding is no longer limited to the introduction of genes from varieties of the same species or sexually compatible species. New sequences can now be obtained from any living organism. It is even possible to design and synthesize them chemically, for example, in order to make them more closely comply with the codon usage of the host plant [56].

Plant transformation not only helps improve plants, it also helps industry. Transgenic cells or plants may be used like factories to over-produce a certain protein that is difficult to extract from normal cells. In addition, animal proteins whose genes have been cloned and expressed in bacteria or even yeast do not always undergo the post-translational modifications (glycosylations, etc.) required for their biological activity. Expressed in plants, they may undergo these modifications, making them look more like native proteins. Plant cells [57] and hairy roots [58] may be grown in perfectly controlled conditions in bioreactors. The proteins produced by transgene expression may be purified relatively easily from the culture medium. To produce recombinant proteins for drug use, these systems have real advantages over the culture of animal cells. For example, animal cells require a more expensive substrate and carry a greater risk of contamination by viruses pathogenic for animals and humans.

These advantages are also found in the cultivation of whole plants. They can use the sun, pure air and fresh water to produce quantities of products with a high added value. The production of *Bacillus licheniformis* α-amylase in transgenic plants allows the direct use of the ground seeds for liquefying starch [59]. Transgenic plants for producing elastomers and biodegradable plastics were obtained by the transfer of chimaeric genes responsible for the biosynthesis of polyhydroxyalcanoates, polyesters that accumulate in the form of inclusions in certain bacteria [60].

Research is currently being carried out on the production of vaccines by plants, either transgenic plants or plants infected with genetically modified viruses so that their capsid presents an epitope of a pathogenic agent for animals or humans [61]. It is easy to obtain large quantities of vaccines prepared from plants. They should be cheaper than those currently prepared from animal or yeast cell cultures. Mice fed with tobacco or potatoes expressing a subunit of a bacterial enterotoxin have been shown to become immune to this toxin [62]. This work will hopefully lead to the development of "edible" vaccines for humans against, for example tooth decay, cholera and AIDS. A&M University in Texas is working on transgenic bananas to protect populations against different infections [63].

The number of antibodies produced by plants has continually increased since 1989 when Hiatt *et al.* described the correct proteolytic maturation and tetrameric assembly of active monoclonal murine antibodies in tobacco plants after crossing two transgenic plants, one of which expressed a heavy chain and the other a light chain [64]. The production of antibodies against phytoviruses provides transgenic plants with partial tolerance [65]. Other functional antibodies have been synthesized in plants, directed against the agent responsible for tooth decay [66,67], the Herpes virus or cancer cells [68]. The large-scale production of recombinant secretory immunoglobulin A may be obtained with transgenic plants and used to develop passive immune therapy. Also, antibodies directed against reactive intermediates (abzymes) shift the equilibrium of reactions and make possible the purification of products that are not normally accumulated.

Methods

Gene transfer involves the introduction of genetic information (in general in the form of DNA) into the genome of a cell. In most cases, the stable expression of this information is required.

Contrary to animal cells, plant cells are surrounded by a rigid pectocellulose wall that prevents the passage of macromolecules such as DNA. In certain plant species, it has been shown, using simple methods, that the cells are totipotent. This is not the case for animal cells. It is possible, for an increasing number of species, by culturing an organ (for example leaf, stem, root) on nutritive media, or by the application of appropriate hormone treatments to obtain the differentiation of the cells and direct their multiplication towards the regeneration of a whole plant. It is therefore potentially possible, if genetic information is integrated in a cell, to obtain a plant where all the cells derived from the manipulated cell possess and express this information. If the regeneration occurs from several cells, a fragment of organ or a callus for example, some of which are transformed, others not, the plant will be chimaeric since all of its cells do not have an identical genome. Since embryos are derived from the fertilization of only one cell, one must wait for the progeny of this primary chimaeric transformant in order to obtain a homogenous genetically modified individual, provided that the germ cells are derived from one of the transformed cells. However, if the regeneration occurs from a single cell, for example a protoplast, the initial transformed plant will have the transgenes in all of its cells.

Transgenic plants obtained by *in vitro* culture may show somaclonal variation. A more or less large proportion of the transformed individuals, according to the cultivar, may present phenotypic variations that are incompatible with their utilization. In general, the only selected individuals are those that do not manifest phenotypic differences with the parent, except for those desired, resulting from the expression of the transgenes.

As mentioned above, two families of techniques are now used for the transformation of plant cells. One involves the use of the properties of *Agrobacterium*, the other uses more or less sophisticated physical or chemical methods that force the penetration of DNA into cells.

Transformation by *Agrobacterium*

The molecular mechanisms underlying the natural genetic transformation by *Agrobacterium* are related to those involved in bacterial conjugation, except that the host is a plant cell [6]. The bacterial chromosome codes for the proteins required for the attachment of the agrobacteria to the plant cells. In addition to T-DNA, the Ti or Ri plasmids carry a so-called virulence region, consisting of many *vir* genes induced by phenolic compounds given off by the wounded plant cells. These *vir* genes code for the proteins responsible for the different steps leading up to the transfer of the T-DNA. The borders of the T-DNA consist of direct repeats of 25 base pairs. The right border is accompanied by a special "overdrive" signal. These borders are recognized and cut on one strand by the action of one of the Vir proteins. A new strand of DNA is synthesized from the gap thereby created in the right border to the gap in the left border. Protected and translocated by proteins that are also coded by *vir* genes, the transferred strand (T-strand) is transported from the bacterial cell to the nucleus of the plant cell. The only conditions needing to be satisfied for a DNA sequence to be transferred to the plant cell are that it has to be between the right and left borders of the T-DNA and the *Agrobacterium* carrying this T-DNA has to have the virulence functions to be able to act in *trans* to carry out the transfer towards the nucleus of the infected plant cells. Usually the T-DNA consists exactly of the region flanked by the direct repetition of 25 base pairs forming the right and left borders recognized by the virulence functions. P. Hooykaas and R. Schilperoot [69] have summarized the steps in this complex mechanism. A simplified diagram is provided in *Figure 1*. Our understanding of the mechanisms of genetic transformation of plants by bacteria is used to transfer the desired genes into plants.

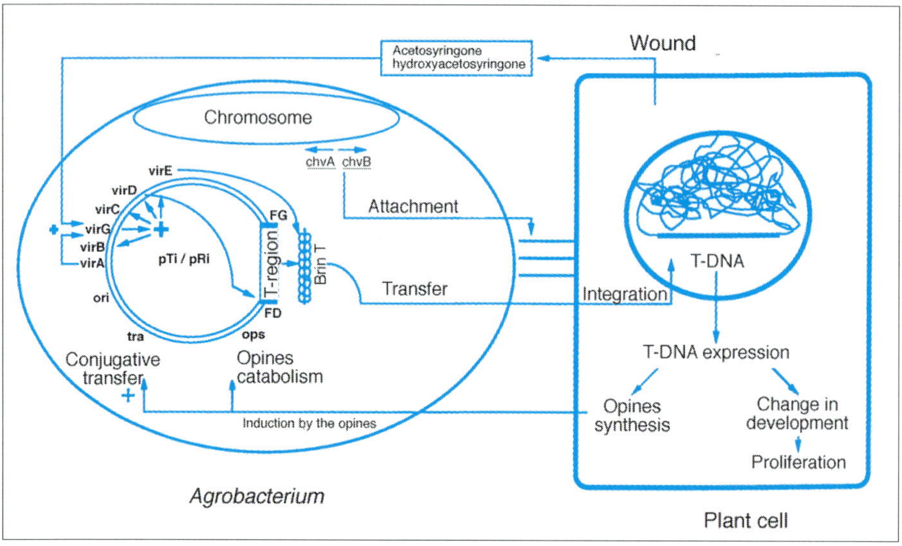

Figure 1. Interaction between *Agrobacterium* and the host plant cell.
Refer to page 59, second paragraph, for an explanation of this diagram.

In the case of *Agrobacterium tumefaciens*, it quickly became evident that it was not possible to regenerate whole plants from transformed cells within the tumour. This is due to the hormone imbalance induced by the expression of some of the T-DNA genes. It is possible, by *in vitro* manipulation of the Ti plasmid, to remove the genes responsible for hormone synthesis from the T-DNA and replace them with a gene providing resistance to a phytotoxic agent (antibiotic or herbicide). During the infection of a tissue fragment from a host plant, the modified ("disarmed") T-DNA retains the ability to be transferred into the nucleus of the peripheral cells of the fragment. However, it no longer induces tumours. If this tissue is cultivated after infection on a medium containing both the selective agent and the growth hormones required for regeneration, only the transformed cells expressing resistance are likely to survive, divide and regenerate whole plants. The untransformed cells will be quickly killed by the selective agent.

In the case of *A. rhizogenes*, the hairy roots may regenerate transformed plants that present a specific phenotype that may be used as a transformation marker [70].

To introduce new genes in plants, the T-region of a disarmed Ti plasmid or Ri plasmid (not necessarily disarmed) is used. The Ti or Ri plasmids include about 200,000 base pairs. Of course, they do not have single restriction sites for the direct cloning of the DNA to be transferred into their T-region. Before being transmitted to *Agrobacterium*, the sequences to be transferred first have to be introduced into a vector that is easily manipulated *in vitro*, *i.e.* it has to be small and able to multiply in *Escherichia coli*. Two types of strategy are used to comply with these requirements:

- Either the foreign gene is introduced into an intermediate vector, a small plasmid that can easily be manipulated *in vitro*, multiplies in *E. coli* and possesses a region of homology with a plasmid bearing the virulence functions. Once transferred into *Agrobacterium* (by conjugation or transformation), this intermediate vector may, by homologous recombination, combine with the Ti or Ri plasmid. The sequences to be transferred to the plants are then found between the left border (LB) and right border (RB) of the T-region of the recombinant plasmid. There are an infinite number of variations on this theme. Depending on the way the intermediate vector and its host plasmid have been designed, the T-DNA that will be transmitted to the plant cell by the cointegrative plasmid will have varying amounts of unnecessary sequence. Three examples are provided in *Figure 2*.

- Or the virulence functions of a helper plasmid are used in *trans* to promote the transfer of a foreign gene introduced into a T-region carried by a compatible independent replicon. Such a binary vector must multiply in *E. coli*, the choice host for *in vitro* manipulation, as well as in *Agrobacterium*, and possesses the whole T-region. An example of a binary system is provided in *Figure 3*. It should be noted that if the vector has only one border, it will act as both left and right borders and the T-DNA will be formed from the whole plasmid.

In order to obtain the expression of new genetic information in transgenic plants, the coding sequence should be placed between a promoter and a termination signal for transcription in plant cells. The cloning site of the intermediate vectors or binary vectors most often used is thereby located between such regulation signals. They are called expression vectors.

One of the first binary vectors was built in 1984 by M. Bevan [71]. The use of this pBin plasmid and its derivatives is so widespread that an American team recently determined the complete sequence [72]. Several DNA fragments of unexplained origin were found to accompany the intended fragments. If it is still necessary, this demonstrates that the molecules obtained by *in vitro* manipulations do not always correspond to those planned "on paper" and that verification by sequencing is always wise.

Whatever the case, once the transfer of chimaeric genes has been obtained, it is possible to obtain transgenic plants *via Agrobacterium* as long as one knows how to select the cells initially transformed by the agrobacteria and regenerate whole organisms from these cells. The efficacy of different selection markers depends on the host plant. However, the selection problem seems to have been resolved in most species. Many antibiotic resistance genes enabling the selection of transformed plant cells have been constructed and introduced into T-DNA. The most common are kanamycin or hygromycin resistance, obtained by the

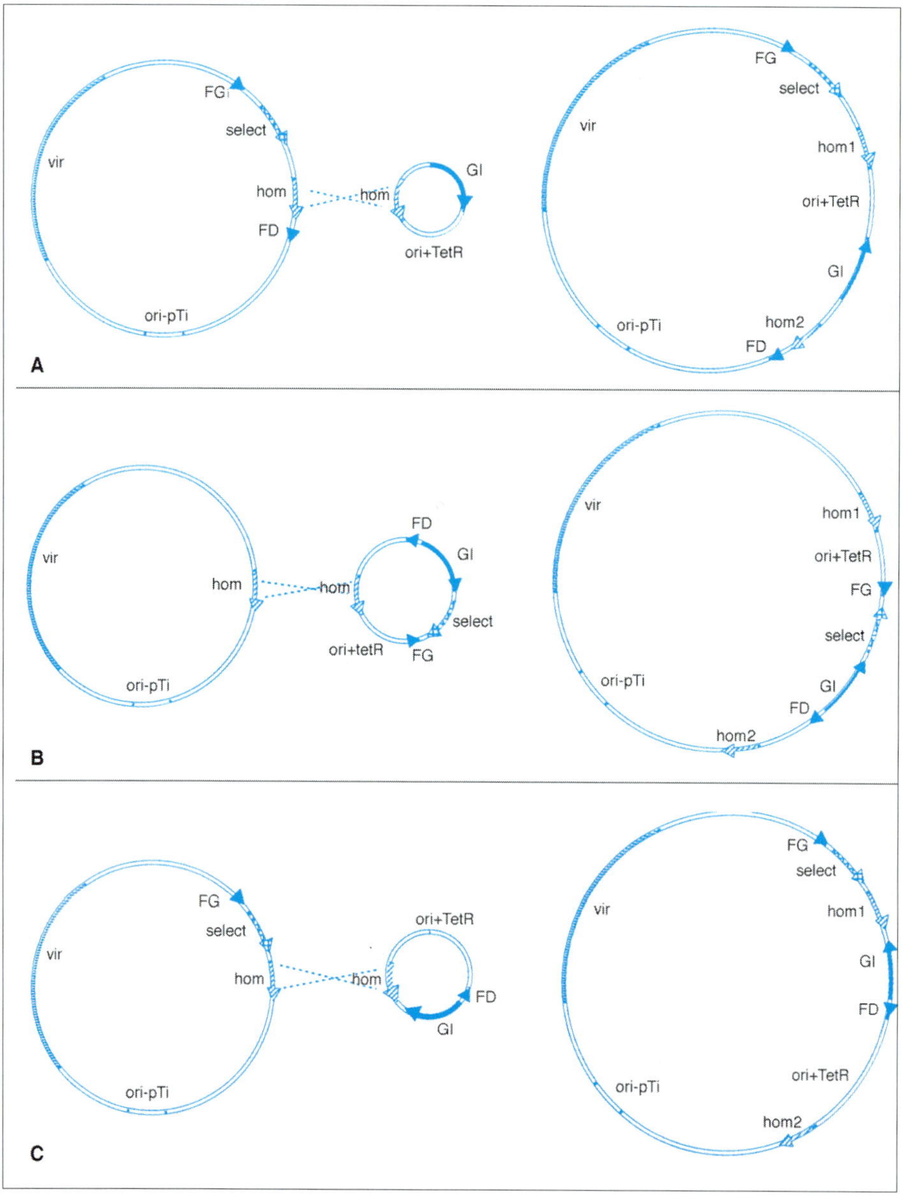

Figure 2. Cointegrative Intermediate Vectors.

A. The gene of interest (*GI*) is cloned in an intermediate vector with, besides a replication origin only fuctional in *E. coli* (ori) and a bacterial selection gene (TetR here), a sequence homologous to part of the T region (hom) making it able to recombine with the T region of a disarmed Ti plasmid carrying the selection marker for the transformed plant cells (*select*). The T-DNA will include the whole sequence of the intermediate vector including the "useless" sequences, *i.e.* the bacterial replication origin (ori) and the bacterial selection marker (TetR).

B. The gene of interest (*GI*) is cloned with the selection marker for the plant cells (*select*) between both boundaries of the T-DNA (LB and RB) carried by an intermediate vector able to combine by homologous recombination with a disarmed Ti plasmid only providing the virulence functions. The T-DNA is limited to the selection marker gene and the gene of interest.
C. The gene of interest (*GI*) is cloned in an intermediate vector possessing the right border of the T- DNA (RB) as well as an origin of replication functional only in *E. coli* (ori), a bacterial selection gene (TetR), and a sequence homologous to part of the T region (hom) making it capable of recombining with a disarmed Ti plasmid bearing the left border (LB) and a selection marker for the transformed plant cells (*select*). In addition to the selection marker gene and the gene of interest, the T-DNA will carry one of the two copies of the fragment that supplied the homology for recombination.

expression of specific phosphotransferases. Herbicide resistance has been developed more recently. Phosphinothricin resistance by expression of the gene coding for a specific acetyl transferase is used more and more often, having the advantage of functioning well in many different host plants.

On the contrary, problems concerning regeneration have been mastered in only a limited (but increasing) number of plant species. The success of transformation still requires the regeneration of the cell into the whole plant and therefore depends on our knowledge of plant biology. Depending on the species to be transformed, *Agrobacterium* infection of protoplasts or stem or leaf fragments may be used. The transgenic plants are selected on regeneration medium containing the selective agent.

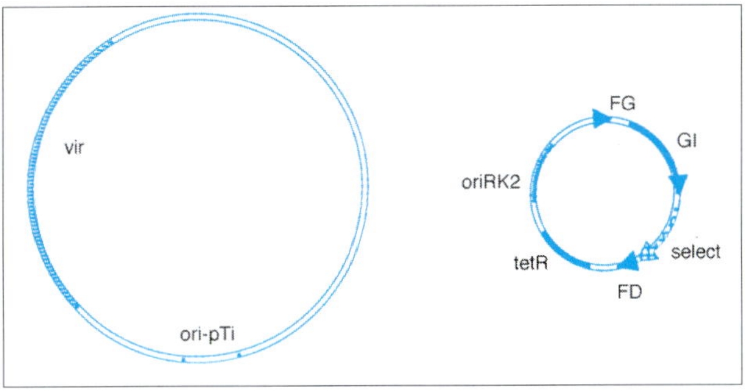

Figure 3. Binary System.

The genes to transfer (selectable marker, select) and the gene of interest (*GI*) are cloned between the boundaries of the T-DNA (RB and LB) in a plasmid possessing an origin of replication functional in *Agrobacterium* and compatible with pTi (here oriRK2), and a bacterial selection marker (here TetR). The helper plasmid will provide the virulence functions able to recognize the boundaries of the T-DNA in *trans*.

In vivo infection of germinating seeds or the stem apex is possible with agrobacteria in certain plants with a specific vegetative system or very small seeds (such as *Arabidopsis thaliana*). Since the organs (embryo or apex) are very small and consist of a small number of cells, certain germline cells are often transformed during the infection. After flowering and self-fertilization, it is possible to obtain seeds with a fully transformed embryo in the progeny since they are derived from a single original egg cell.

Transgenic plants have been obtained in many species by using disarmed Ti plasmids or Ri plasmids. For a long time, most of foreign genes were transferred to tobacco for studying the regulation of expression since tobacco is easily manipulated *in vitro*. Since transformation of *Arabidopsis* has been mastered, this plant has become the model and is slowly but surely replacing tobacco. Transgenic potato, tomato, lettuce, rape, cabbage, cotton, soybean and alfalfa are now also obtained with *Agrobacterium* even though the techniques used are not effective on all of the varieties. The tendency to regenerate not only depends on the species and cultivar, it also depends on the type and age of the tissue involved in the primary transformation. Certain species or varieties of commercial interest are recalcitrant in providing transgenic individuals after transformation by *Agrobacterium*. However, this difficulty will undoubtedly be overcome by further research.

Transformation most often involves disarmed vectors. However, in certain cases, the specific phenotype of the plants transformed by *A. rhizogenes*, derived from hairy root culture, may be used to screen or select transgenic plants. An intermediate vector that co-integrates with a pRi may be used.

If we want to eliminate the selection marker (resistance or hairy root phenotype) from the progeny of transgenic plants and only keep the desirable gene, a strain of *Agrobacterium* may be used that has two plasmids, one with the virulence functions and T-DNA containing the selection marker, the other a binary vector carrying T-DNA with only the desired gene. The transfer of the two types of T-DNA is possible due to the virulence functions of the former plasmid (*Figure 4*). Two bacteria can also be used, one transferring the selection marker (or carrying a wild pRi), the other with a binary system with disarmed assistant plasmid. A large percentage of the cells that receive T-DNA carrying the selection marker also acquire the T-DNA carrying the desirable gene. Since each T-DNA comes from a different bacteria, the two transformation events are independent and the two sites of chromosome insertion may be far apart, thereby enabling their segregation during meiosis.

Most dicots are sensitive to *Agrobacterium*. Besides the species mentioned above, the regeneration of transgenic plants is possible from woody species such as rose, vine, poplar, beech, citrus, eucalyptus, apple, peach and plum. For a long time, the transformation of monocots by *Agrobacterium* was thought to be impossible since they are not susceptible to crown gall or hairy root. In fact, only the problem of

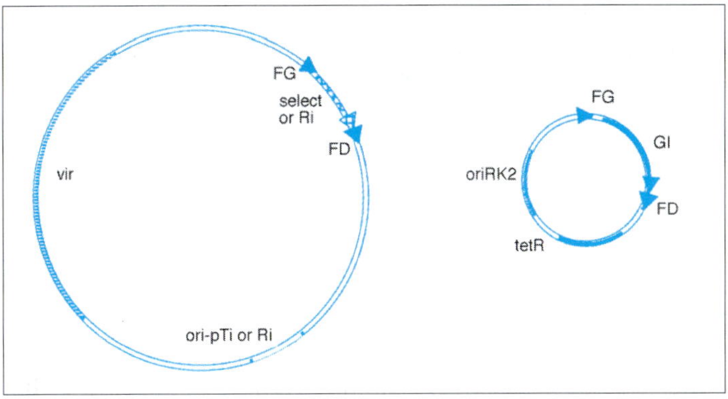

Figure 4. Binary system with selectable marker physically separate from the desirable gene. Use of plamid bearing T-DNA with the gene of interest (*GI*) operating in a binary system either with a wild Ri plasmid or a virulence plasmid containing T-DNA with a chimaeric resistance gene (*select*). The 2 T-DNAs will often be co-transferred and they may segregate in the progeny if they are not linked in the nucleus of the transgenic plant obtained.

regeneration from isolated cells seems to prevent the use of this transformation method. Some time ago, viral DNA cloned between T-DNA borders was introduced by *Agrobacterium* into maize and wheat cells resulting in a systemic viral infection (agro-infection) [73]. Recently, using a specific strain of *Agrobacterium*, it has been possible to obtain transgenic rice from callus derived from the scutella of different cultivars of Japonica rice [74].

The difficulties in generalizing the use of *Agrobacterium* for obtaining transgenic plants mainly stem from problems in the choice of strains to use to provide the virulence functions and the choice of target cells to initially transform in order to obtain regeneration.

Direct Gene Transfer

Parallel to the development of transformation techniques by *Agrobacterium*, other transformation strategies have been considered. They aim at the direct transfer of genes, without a bacterial intermediate, into plant cells.

Contrary to techniques derived from the transformation by *Agrobacterium*, which require that the DNA to be transferred be bordered with well determined nucleotide sequences, these techniques do not require the use of specific vectors. The genetic information has to be amplified, *i.e.* cloned, for a sufficient quantity to be available for its purification and transfer. It is obtained in the form of a plasmid multiplied in *E. coli*. This assumes the presence of at least one replication origin and one selection marker functional in the bacteria (resistance to an antibiotic). Classic vectors from the pUC series are often used. The genes to be transferred must have a eukaryotic structure to be expressed in plant cells. If the expression of the desirable gene does not produce a selectable phenotype, it is necessary to co-transfer a selection marker, as is true in transformation by *Agrobacterium*.

The preparation of the DNA for transfer should include both the gene enabling the selection of the transformed plant cells and the desirable gene. Either these two genes are carried by the same vector or a mixture of two plasmids are co-transferred, each carrying one of the genes, the proportions being that any plant cell that receives the selection gene also has a chance of receiving the desirable gene. The prepared DNA, in native circular form or linearized by the action of restriction enzymes, may then be directly transferred to the protoplasts or cells, tissues or organs. The first direct transfer experiments used carrier DNA (from calf thymus for example) to increase the efficiency of the transfer. As pointed out by the Biomolecular Engineering Commission, in such cases, it is impossible to identify the sequences of carrier DNA transferred. This strategy has been abandoned little by little. The techniques used in the preparation of plasmid DNA have become simpler, the methods of transformation have improved and carrier DNA is no longer indispensable.

Protoplast Transformation

Protoplasts are plant cells from which the pectocellulose cell wall has been removed by enzyme treatment. They can be obtained from different organs in all plants. It is possible to carry out a full cultivation cycle with certain species, *i.e.* it is possible to regenerate whole plants from these protoplasts after cultivation *in vitro*, although the efficiency varies.

The transformation techniques developed for use with animal cells can be successfully applied to these protoplasts, where the cytoplasm is only isolated from the external environment by a plasma membrane. Polyethylene glycol (PEG) was first used. This non-toxic compound is able to destabilize the plasma membrane and allows the DNA to be transferred through the membrane [75]. The DNA molecules can then migrate to the nucleus. Some of them are able to combine with the chromosomes with variable success. If this genetic information is correctly expressed and confers the protoplast with a selectable character (such as antibiotic or herbicide resistance, for example), it is possible, after growth of the treated protoplasts on a selective medium, to regenerate transformed plants expressing this new character.

Another protoplast transformation technique was developed by provoking fusion, in the presence of PEG, between the protoplasts and liposomes, (small artificial phospholipid vesicles) containing the DNA to be transferred. The similarity of the membranes of protoplasts and liposomes allows them to merge. The liposomes then empty their contents into the cytoplasm of the protoplasts [76].

The most recent and one of the most effective techniques is also one of the simplest to use: protoplast electroporation [77]. A mixture of protoplasts and DNA are subjected to a series of short and intense electric shocks. The electric field destabilizes the plasma membrane by polarization of the phospholipids forming it. This forms pores through which the DNA molecules can transit. If the electric shock is not too violent, the phenomenon is reversible and the membrane returns to its initial state, leaving the protoplast perfectly viable.

The techniques using plant protoplasts are fairly difficult to apply and cannot be applied to many species where plant regeneration from protoplasts is still not mastered. However, these techniques have been used to transform important cereals such as rice [78], maize [79] and barley [80].

Direct Transformation of Cells, Tissues or Organs

These techniques involve the penetration of the DNA through the pectocellulose wall of plant cells. At the present time, particle bombardment is the only technique that reproducibly works [81]. The principle consists in bombarding the tissue with DNA-coated small gold or tungsten particles. Many types of bombardment have been developed that use different types of particle propulsion. The explosion of a powder cartridge, the excess pressure provoked by the instant vaporization of a drop of water or the expansion of a compressed gas can be used to project the particles onto the tissue in a violent but controlled manner under reduced air pressure. They have enough kinetic energy to cross the cell membrane, however, their small size does not cause any irreparable damage. Once in the cytoplasm, the DNA may solubilize and migrate to the nucleus where it can be transiently expressed or combine stably with a chromosome.

This technique is relatively difficult to apply although it has several advantages over the protoplast methods. Although it is still necessary to regenerate the plants after the selection of the transformed cells, it is not necessary to be able to isolate and culture protoplasts since it is possible to work directly on cells or tissues in culture. Finally, it is sometimes not even necessary to know how to grow the plant tissues *in vitro*. It is possible to bombard the apex or embryos directly. In this case, a chimaeric plant, a mosaic of transformed and untransformed cells, is obtained after the development of the treated organ. If certain cells at the origin of the germ line are transformed, this chimaeric plant will produce seeds with entirely transformed embryos. The method is long and difficult. However, the feasibility has been proven as indicated by the success of the experiments carried out on many species such as soybean [82], cotton [83] and cereals such as maize [84], wheat [85,86], oats [87], rice [88], rye [89], barley [90] and sorghum [91].

Many other direct transfer techniques are being studied, such as transformation by pollen, tissue sonication, embryo imbibition, or the macroinjection of DNA into tissues. However, the results have only been sporadic and difficult to reproduce. Recently, certain companies have obtained excellent results with carbon fibres carrying adsorbed DNA.

Transformation of Organelles

Until now, the transgenic plants tested in the field are all derived from the transformation of their nuclear genome. Their transgenes are incorporated into chromosomes and follow Mendelian inheritance. However, it should be noted that direct transfer methods have also made it possible to introduce new genetic information into the genomes of chloroplasts or mitochondria, cell organelles that are mostly maternally inherited. Although most of the proteins found in these organelles are coded by the nuclear genome, some of their major functions depend on their own genome. It is therefore obvious that, over the next few years, we will see the further development of these approaches and probably their application to the modification of characteristics related to the efficiency of respiration or photosynthesis. It is now possible to transform the chloroplast genome and obtain integration by homologous or targeted recombination, something that is still difficult with the nuclear genome [92,93].

Viruses as Vectors

Although this does not strictly speaking involve transgenic plants, it should be kept in mind that viruses have a huge potential to express proteins. This is beginning to be used in the production of foreign proteins in infected plants. These obligatory parasites have a relatively simple genome and may be manipulated *in vitro*. Part of their genetic information can be replaced by that directing the synthesis of the desirable protein(s) [94]. The vectors may be derived from a DNA virus or an RNA virus and, in certain cases, be transmitted to the plant by agro-infection. The infection of a plant by such recombinant viruses will lead to their systemic propagation and thereby the high-level expression of the desired genetic information in the whole plant [95]. Viral vectors have already been used to produce a protein able to inhibit the replication of HIV [96] or vaccines [61].

One prospect opened up by the use of viruses as vectors lies in the fact that it is becoming possible to obtain the expression of new genes during the life of the plant. This will allow for the improvement of the characteristics of perennial species without having to regenerate new genetically modified individuals.

Molecular Characterization of Transformants

Whatever the transformation technique used, the next step involves the molecular characterization of the transgenic plants selected. It is important to know the number of copies of the transgene present and/or the number of transgenes expressed and also whether rearranged or silent (non-expressed) copies are also present. It is important to know the type and extent of the DNA that may have been transferred outside of the desirable sequences. How is it possible to give an exact description of the molecular events related to the integration of the transgenes?

In the case of transformation by *Agrobacterium*, it is first necessary to check that the bacteria are no longer found within the plant tissue. This requires Southern analysis to demonstrate the absence of sequences corresponding to the virulence genes, for example, or any other bacterial nucleotide sequences normally not transferred. As noted above, the only anticipated transferred sequence is that contained within the right and left borders of the T-DNA. The Biomolecular Engineering Commission has always asked applicants not to rely solely on the bibliographical data on the transfer mechanism *via Agrobacterium* but to verify that it has functioned as expected. Many transformants have been abandoned as the left border had been overrun. Calgene, benefiting from experience with several hundred transformants belonging to different plant species, has observed that the transferred region overlaps the left border in 20% to 30% of the transgenic plants obtained with transformation by *Agrobacterium* [97]. It is therefore necessary to check that the T-DNA boundaries have been respected, and especially that sequences such as a wide spectrum replication origin or bacterial antibiotic resistance genes have not been integrated into the genome of the selected plants. These sequences would not only be useless but also a potential risk, if one considers the possible passage of genetic information from higher plants to the microbe flora in the soil, however improbable.

In the case of direct DNA transfer, if the transformation is carried out with a circular plasmid, its integration is expected in a linear form in an uncertain manner at any site whatsoever. The direct transfer of DNA may lead to the insertion of altered or inactive copies of the sequences introduced. Only molecular analysis can reveal this.

Two types of analysis are used to determine the extent and nature of the sequences transferred: Southern transfer and hybridization or amplification by PCR (*Figure 5*). Southern analysis detects sequences homologous to those derived from the plasmids used as probes in the DNA of the transgenic plants. The use of different restriction enzymes to cut the DNA studied is used to map the insert. A sequence that cannot hybridize with the DNA of transgenic plants was not transferred. However, the most exact results are obtained by polymerization chain reaction (PCR). Starting with the primers on the edge of a region that is supposedly transferred intact, a fragment of the expected size with the same restriction sites

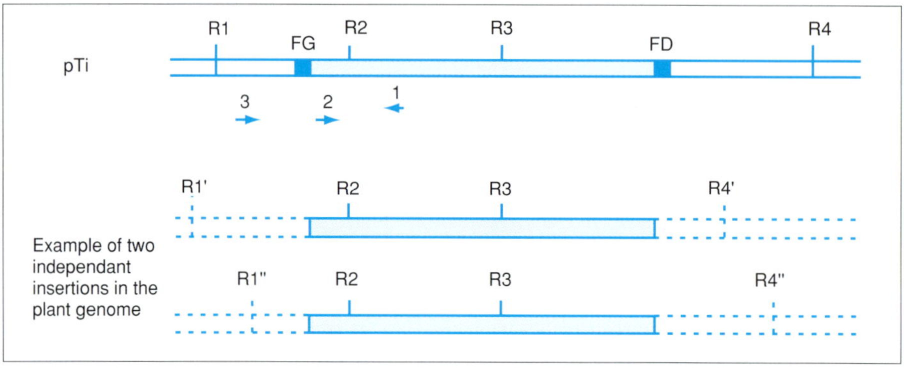

Figure 5. Analysis of the Transferred DNA.

With the Southern method, a probe consisting of the internal restriction fragment of the T-DNA, between sites R2 and R3, will reveal the same fragment in pTi DNA and in DNA from the transformed cells. Both are digested by the enzyme R. However, a probe consisting of the fragment delimited by sites R1 and R2 will hybridize with the plant DNA but the size of the fragment revealed will depend on the position of the closest restriction site cut by the same enzyme (R1' or R1") in the genome. In addition, if there are multiple insertions, it is possible to obtain many such boundary fragments. With the use of several restriction enzymes and a set of appropriate probes, this method can be used to estimate the size and minimum number of copies of T-DNA in the genomic DNA of the transgenic plant. By PCR analysis, amplification with primers 1 and 2 should provide an identical fragment including site R2 when using the plant DNA as well as pTi. However, amplification with primers 1 and 3 should provide a fragment only with pTi if the DNA beyond the left boundary was not transferred to the plant genome.

has to be amplified. The amplification of the vector and that of the transgenic plant DNA should give exactly the same fragment. The use of primers defining a region found on the vector but not transferred should amplify a fragment from the DNA of the vector but not from the genomic DNA of the transgenic plant. Finally, if the primers flank a sequence that was transferred but whose size or structure was altered in the genome of the analysed plant, the amplifications from plasmid DNA or genomic plant DNA will provide fragments of different size or structure.

Full analysis of genetically modified plants also involves the study of the RNA transcribed from the transferred DNA and proteins produced by the translation of this RNA. Molecular probing on the RNA extracted from the transgenic plants (northern blots) and measurements of enzyme activity or their immunodetection (western blots) are carried out.

Conclusion and Prospects for Plant Transformation

The success of transformation for introducing desired traits into commercial varieties indicates that a wish to transfer new transgenes into already modified plants will soon arise. Since the number of selectable genes is limited, the problem of the elimination of the marker that was used to select the transformed individuals will become of increasing concern. These markers are no longer of use once the desired individuals have been obtained. Different systems have been used [98]: co-transformation followed by the segregation of the two transgenes has already been mentioned above; site-specific recombination systems such as the cre-lox system of bacteriophage P1 have turned out to be functional in plants and can be used to eliminate a selection marker [99]; finally, transposable elements may be manipulated in order to reposition the transgene within the modified genome [100]. Replacement of the gene by homologous recombination would be the ideal method to target the insertion of transgenes. However, further study is required before this dream, already reality in animal cells, will be fulfilled in plant cell nuclei.

Although most of the transgenic plants undergoing field tests in Europe and in the United-States, except for cotton and rice, are mainly large-scale crop species in temperate climates, the development of plant transformation does not exclusively concern the industrialized countries. Research is being carried out on a certain number of tropical plants. The methods of transformation have been described, sometimes with the regeneration of transgenic plants. Tests have been carried out on virus-resistant papaya plants over the last few years [101,102]. During the next few years, transgenic plants belonging to many other tropical species will probably be developed. Studies have been carried out by ORSTOM researchers in order to obtain the transformation of cassava. The use of *A. rhizogenes* gave rise to transgenic coffee-shrubs [103] while particle bombardment has transformed sorghum [91] and sugar cane [104]. Genetically modified banana trees have been obtained *via Agrobacterium* [105] and particle bombardment [106]. The genetic transformation of cocoa leaf cells was obtained *via Agrobacterium tumefaciens* [107] and in rubber callus after particle bombardment [108]. The efforts devoted to obtain regeneration will surely be soon successful.

The first transformed plants were obtained only about twelve years ago [7]. During this period, many transformation techniques have been developed, each with different characteristics and efficiency. The progress made shows that major emphasis is put on simplifying the methods used to obtain transgenic plants. As J. Schell stated in June 1994 during an international symposium on plant molecular biology in Amsterdam, the term "refractory to transformation", is little by little becoming obsolete [109]. The effort made to develop all-purpose, simple methods will intensify the use of genetic transformation. It is a wonderful tool to study plant biology as well as an additional instrument in the hands of the plant breeder.

Although a list of genes introduced and species transformed would seem long, the development of plant transformation is only in its initial stages and any attempt to describe the state of the art will be obsolete before publication. Although the enthusiasm of researchers in this rapidly expanding field is positive and should be encouraged, precautions are still required concerning transformation, as with any emerging technique, if we want to greet the 21st century, not as "mad scientists" but as fully responsible civilized beings.

References

1. Goldman A, Tempé J, Morel G. Quelques particularités de diverses souches d'*Agrobacterium tumefaciens*. *CR Soc Biol* 1968 ; 162 : 630-1.
2. Van Larebeke N, Engler G, Holsters M, Van den Elsacker S, Zaenen I, Schilperoort RA, Schell J. Large plasmid in *Agrobacterium tumefaciens* essential for crown gall-inducing ability. *Nature* 1974 ; 252 : 169-70.
3. Zaenen I, Van Larebeke N, Teuchy H, Van Montagu M, Schell J. Supercoiled circular DNA in crown gall inducing *Agrobacterium* strains. *J Mol Biol* 1974 ; 86 : 109-27.
4. Chilton MD, Drummond MH, Merlo DJ, Sciaky D, Montoya AL, Gordon MP, Nester EW. Stable incorporation of plasmid DNA into higher plant cells: the molecular basis of crown gall tumorigenesis. *Cell* 1977 ; 11 : 263-71.
5. Chilton MD, Tepfer DA, Petit A, David C, Casse-Delbart F, Tempé J. *Agrobacterium rhizogenes* inserts T-DNA into the genomes of the host-plant root cells. *Nature* 1982 ; 295 : 432-4.
6. Stachel SE, Zambryski P. *Agrobacterium tumefaciens* and the susceptible plant cell: a novel adaptation of extracellular recognition and DNA conjugation. *Cell* 1986 ; 47 : 155-7.
7. Herrera-Estrella L, Deblock M, Messens E, Hernalsteens JP, Van Montagu M, Schell J. Chimeric genes as dominant selectable markers in plant cells. *EMBO J* 1983 ; 2 : 987-95.
8. Zambryski P, Joss H, Genetello C, Leemans J, Van Montagu M, Schell J. Ti plasmid vector for the introduction of DNA into plant cells without alteration of their normal regeneration capacity. *EMBO J* 1983 ; 2 : 2143-50.
9. Schell J. Transgenic plants as tools to study the molecular organization of plant genes. *Science* 1987 ; 237 : 1176-83.
10. Kuhlemeier C, Green PJ, Chua NH. Regulation of gene expression in higher plants. *Annu Rev Plant Physiol* 1987 ; 38 : 221-57.
11. Van der Krol A, Mur LA, de Lange P, Mol NM, Stuitje AR. Inhibition of flower pigmentation by antisense CHS genes: promoter and minimal sequence requirements for the antisense effect. *Plant Mol Biol* 1990 ; 14 : 457-66.
12. Matzke MA, Primig M, Trnovsky J, Matzke AJM. Reversible methylation and inactivation of marker genes in sequentially transformed tobacco plants. *EMBO J* 1989 ; 8 : 643-9.
13. Dorlhac de Borne F, Vincentz M, Chupeau Y, Vaucheret H. Co-suppression of nitrate reductase host genes and transgenes in transgenic tobacco plants. *Mol Gen Genet* 1994 ; 243 : 613-21.
14. Feldmann KA. T-DNA insertion mutagenesis in *Arabidopsis*: mutational spectrum. *Plant J* 1991 ; 1 : 71-82.
15. Walden R, Hayashi H, Schell J. T-DNA as a gene tag. *Plant J* 1991 ; 1 : 281-8.
16. Bechtold N, Ellis J, Pelletier G. Transformation *in planta* de plantes adultes d'*Arabidopsis thaliana* par infiltration d'*Agrobacterium*. *CR Acad Sci Paris* 1993 ; 316 : 1194-9.
17. Vaeck M, Reynaerts A, Höfte H, Jansens S, De Beuckeleer M, Dean C, Zabeau M, Van Montagu M, Leemans J. Transgenic plants protected from insect attack. *Nature* 1987 ; 327 : 33-7.
18. Barton KA, Whiteley HR, Yang NS. *Bacillus thuringiensis* δ-endotoxin expressed in transgenic *Nicotiana tabacum* provides resistance to lepidopteran insects. *Plant Physiol* 1987 ; 85 : 1103-9.

19. Fischhoff DA, Bowdish KS, Perlak FJ, Marrone PG, McCormick SM, Niedermeyer JG, Dean DA, Kusano-Kretzmer K, Mayer EJ, Rochester DE, Rogers SG, Fraley RT. Insect tolerant transgenic tomato plants. *Bio/Technology* 1987 ; 5 : 807-13.
20. Hilder VA, Gatehouse AMR, Sheerman SE, Barker RF, Boulter D. A novel mechanism of insect resistance engineered into tobacco. *Nature* 1987 ; 300 : 160-3.
21. Johnson R, Narvaez J, An G, Ryan C. Expression of proteinase inhibitors I and II in transgenic tobacco plants: effects on natural defense against *Manduca sexta* larvae. *Proc Natl Acad Sci USA* 1989 ; 86 : 9871-5.
22. Shade RE, Schroeder HE, Pueyo JJ, Tabe LM, Murdock LL, Higgins TJV, Chrispeels MJ. Transgenic pea seeds expressing the α-amylase inhibitor of the common bean are resistant to bruchid beetles. *Bio/Technology* 1994 ; 12 : 793-6.
23. Sanford JC, Johnston SA. The concept of parasite-derived resistance. Deriving resistance genes from the parasite's own genome. *J Theor Biol* 1985 ; 113 : 395-405.
24. Powell Abel P, Nelson RS, De B, Hoffmann N, Rogers SG, Fraley RT, Beachy RN. Delay of desease development in transgenic plants that express the tobacco mosaic virus coat protein gene. *Science* 1986 ; 232 : 738-43.
25. Baulcombe DC. Replicase-mediated resistance: a novel type of virus resistance in transgenic plants? *Trends Microbiol* 1994 ; 2 : 60-3.
26. Brederode FT, Taschner PEM, Posthumus E, Bol JF. Replicase-mediated resistance to alfalfa mosaic virus. *Virology* 1995 ; 207 : 467-74.
27. Lapidot M, Gafny R, Ding B, Wolf S, Beachy RN. A dysfunctional movement protein of tobacco mosaic virus that partially modifies the plasmodesmata and limits virus spread in transgenic plants. *Plant J* 1995 ; 4 : 959-70.
28. Lamb CJ, Ryals JA, Ward ER, Dixon RA. Emerging strategies for enhancing crop resistance to microbial pathogens. *Bio/Technology* 1992 ; 10 : 1436-45.
29. Broglie K, Chet I, Holliday M, Cressman R, Biddle P, Knowlton S, Mauvais CJ, Broglie R. Transgenic plants with enhanced resistance to the fungal pathogen *Rhizoctonia solani*. *Science* 1991 ; 254 : 1194-7.
30. Zhu Q, Maher EA, Masoud S, Dixon R, Lamb CJ. Enhanced protection against fungal attack by constitutive co-expression of chitinase and glucanase genes in transgenic tobacco. *Bio/Technology* 1994 ; 12 : 807-12.
31. Logemann J, Jach G, Tommerup H, Mundy J, Schell J. Expression of a barley ribosome-inactivating protein leads to increased fungal protection in transgenic tobacco plants. *Bio/Technology* 1992 ; 10 : 305-14.
32. Hain R, Reif HJ, Krause E, Langebartels R, Kindl H, Vorman B, Wiese W, Schmellzer E, Schreirer PH, Stöcker RH, Stenzel K. Disease resistance results from foreign phytoalexin expression in a novel plant. *Nature* 1993 ; 361 : 153-6.
33. Fischer R, Hain R. Plant disease resistance resulting from the expression of foreign phytoalexins. *Curr Opin Biotechnol* 1994 ; 5 : 125-30.
34. Düring K, Porsch P, Fladung M, Lörtz H. Transgenic potato plants resistant to the phytopathogenic bacterium *Erwinia carotovora*. *Plant J* 1993 ; 3 : 587-98.
35. Jaynes JM, Xanthopoulos KG, Destéfano-Beltran L, Dodds JH. Increasing bacterial disease resistance in plants utilizing antibacterial genes from insects. *BioEssays* 1987 ; 6 : 263-70.
36. Jaynes JM, Nagpala P, Destéfano-Beltran L, Huang JH, Kim JH, Denny T, Cetiner S. Expression of a cecropin B lytic peptide analog in transgenic tobacco confers enhanced resistance to bacterial wilt caused by *Pseudomonas solanacearum*. *Plant Sci* 1993 ; 89 : 43-53.
37. Guerche P, De Almeida ERP, Schwarztein MA, Gander E, Krebbers E, Pelletier G. Expression of the 2S albumin from *Bertholletia excelsa* in *Brassica napus*. *Mol Gen Genet* 1990 ; 221 : 306-14.

38. Pen J, Verwoerd TC, van Paridon PA, Beudeker RF, van den Elzen PJM, Geerse K, van der Klis JD, Versteegh HAJ, van Ooyen AJJ, Hoekema A. Phytase-containing transgenic seeds as a novel feed additive for improved phosphorus utilization. *Bio/Technology* 1993 ; 11 : 811-4.
39. Visser RGF, Jacobsen E. Towards modifying plants for altered starch content and composition. *Trends Biotechnol* 1993 ; 11 : 63-8.
40. Witty M, Harvey WJ. Sensory evaluation of transgenic *Solanum tuberosum* producing r-thaumatin II. *NZ J Crop Horticul Sci* 1990 ; 18 : 77-80.
41 Peñarrubia L, Kim R, Giovannoni J, Kim SH, Fischer RL. Production of the sweet protein monellin in transgenic plants. *Bio/Technology* 1992 ; 10 : 561-4.
42. Töpfer R, Martini N, Schell J. Modification of plant lipid synthesis. *Science* 1995 ; 268 : 681-6.
43. Ohlrogge JB. Design of new plant products: engineering of fatty acid metabolism. *Plant Physiol* 1994 ; 104 : 821-6.
44. Kinney AJ. Genetic modification of the storage lipids of plants. *Curr Opin Biotechnol* 1994 ; 5 : 144-51.
45. Mol JNM, Stuitje AR, van der Krol A. Genetic manipulation of floral pigmentation genes. *Plant Mol Biol* 1989 ; 13 : 287-94.
46. Napoli C, Lemieux C, Jorgensen R. Introduction of a chimeric chalcone synthase gene into petunia results in reversible co-soppression of homologous genes in *trans*. *Plant Cell* 1990 ; 2 : 279-89.
47. Van der Krol AR, Mur LA, Beld M, Mol JNM, Stuitje AR. Flavonoid genes in Petunia: addition of a limited number of gene copies may lead to a suppression of gene expression. *Plant Cell* 1990 ; 2 : 291-9.
48. Kramer M, Sheehy RE, Hiatt WR. Progress towards the genetic engineering of tomato fruit softening. *Trends Biotechnol* 1989 ; 7 : 191-4.
49. Sheehy RE, Kramer M, Hiatt WR. Reduction of polygalacturonase activity in tomato fruit by antisense RAN. *Proc Natl Acad Sci USA* 1988 ; 85 : 8805-9.
50. Tieman DM, Harriman RW, Ramanohan G, Handa AK. An antisense pectin methylesterase gene alters pectin chemistry and soluble solids in tomato fruit. *Plant Cell* 1992 ; 4 : 667-79.
51. Hamilton AJ, Lycett GW, Grierson D. Antisense gene that inhibits synthesis of the hormone ethylene in transgenic plants. *Nature* 1990 ; 346 : 284-8.
52. Oeller PW, Min-Wong L, Taylor LP, Pike DA, Theologis A. Reversible inhibition of tomato fruit senescence by antisense RNA. *Science* 1991 ; 254 : 437-9.
53. Martineau B, Summerfelt KR, Adams DF, DeVerma JW. Production of high solid tomatoes through molecular modification of levels of the plant growth regulator cytokinin. *Bio/Technology* 1995 ; 13 : 250-4.
54. Mariani C, Gossele V, De Beuckeleer M, De Block M, Goldberg RB, De Greef W, Leemans J. Induction of male sterility in plants by a chimaeric ribonuclease gene. *Nature* 1990 ; 347 : 737-41.
55. Mariani C, Gossele V, De Beuckeleer M, De Block M, Goldberg R, De Greef W, Leemans J. A chimaeric ribonuclease-inhibitor gene restores fertility to male sterile plants. *Nature* 1992 ; 357 : 384-7.
56. Perlak FJ, Fuchs RL, Dean DA, McPherson SL, Fischhoff DA. Modification of the coding sequence enhances plant expression of insect control protein genes. *Proc Natl Acad Sci USA* 1991 ; 88 : 3324-8.
57. Westphal K. Large scale production of new biologically active compounds in plant cell cultures. In : Nijkamp HJ, Van der Plas LHW, Van Aartrijk KJ, eds. *Progress in plant cellular and molecular biology*. Dordrecht : Kluwer Publishers, 1990 : 601-8.
58. Wilson PDG, Hilton MG, Robins RJ, Rhodes MJC. The cultivation of transformed roots from laboratory to pilot plant. In : Nijkamp HJ, Van der Plas LHW, Van Aartrijk KJ, eds. *Progress in plant cellular and molecular biology*. Dordrecht : Kluwer Publishers, 1990 : 700-5.

59. Pen J, Molendjik L, Quax WJ, S PC, van Ooyen AJJ, van den Elzen PJM, Rietveld K, Hoekema A. Production of active *Bacillus licheniformis* alpha-amylase in tobacco and its application in starch liquefaction. *Bio/Technology* 1992 ; 10 : 292-6.
60. Poirier Y, Nawrath C, Somerville C. Production of polyhydroxyalkanoates, a family of biodegradable plastics and elastomers, in bacteria and plants. *Bio/Technology* 1995 ; 13 : 142.
61. Turpen TH, Reinl SJ, Charoenvit Y, Hoffman SL, Fallarmo V, Grill LK. Malarial epitopes expressed on the surface of recombinant Tobacco Mosaic Virus. *Bio/Technology* 1995 ; 13.
62. Haq TA, Mason HS, Clements JD, A CJ. Oral immunization with a recombinant bacterial antigen produced in transgenic plants. *Science* 1995 ; 268 : 714-6.
63. Moffat AS. Exploring transgenic plants as a new vaccine source. *Science* 1995 ; 268 : 659-60.
64. Hiatt A, Cafferkey R, Bowdish K. Production of antibodies in transgenic plants. *Nature* 1989 ; 342 : 76-8.
65. Tavladoraki P, Benvenuto E, Trinca S, Martinis DD, Cattaneo A, Galeffi P. Transgenic plants expressing a functional single-chain Fv antibody are specifically protected from virus attack. *Nature* 1993 ; 366 : 469-72.
66. Hiatt A, Ma JKC. Monoclonal antibody engineering in plants. *FEBS Lett* 1992 ; 307 : 71-5.
67. Ma JKC, Leher T, Stabila P, Fux I, Hiatt A. Assembly of monoclonal antibodies with IgG1 and IgGA heavy chain domains in transgenic tobacco plants. *Eur J Immunol* 1994 ; 24 : 131-8.
68. Russel D. *Production of industrial and biopharmaceutical proteins in transgenic plants.* International Meeting of Production of Recombinant Proteins in Plants. Leicester, 1994 : 43.
69. Hooykaas PJJ, Schilperoort R. *Agrobacterium* and plant genetic engineering. *Plant Mol Biol* 1992 ; 19 : 15-38.
70. Tepfer D. Transformation of several species of higher plants by *Agrobacterium rhizogenes*: sexual transmission of the transformed genotype and phenotype. *Cell* 1984 ; 37 : 959-67.
71. Bevan M. Binary *Agrobacterium* vectors for plant transformation. *Nucleic Acids Res* 1984 ; 12 : 8711-21.
72. Frisch DA, Harris-Haller LW, Yokubaitis NT, Thomas TL, Hardin SH, Hall TC. Complete sequence of the binary vector Bin 19. *Plant Mol Biol* 1995 ; 27 : 405-9.
73. Grimsley N, Hohn T, Davies JW, Hohn B. *Agrobacterium*-mediated delivery of infectious maize streak virus into maize plants. *Nature* 1987 ; 325 : 177-9.
74. Hiei Y, Ohta S, Komari T, Kumashiro T. Efficient transformation of rice (*Oriza sativa L.*) mediated by *Agrobacterium* and sequence analysis of the boundaries of the T-DNA. *Plant J* 1994 ; 6 : 271-82.
75. Paszkowski J, Shillito RD, Saul M, Mandak V, Hohn T, Hohn B, Potrykus I. Direct gene transfer to plants. *EMBO J* 1984 ; 3 : 2717-22.
76. Deshayes A, Herrera-Estrella L, Caboche M. Liposome mediated transformation of tobacco mesophyll protoplasts by an *Escherichia coli* plasmid. *EMBO J* 1985 ; 4 : 2731-7.
77. Fromm M, Taylor LP, Walbot V. Expression of genes transferred into monocot and dicot plant cells by electroporation. *Proc Natl Acad Sci USA* 1985 ; 82 : 5824-8.
78. Toriyama K, Arimoto Y, Uchimiya H, Hinata K. *Bio/Technology* 1988 ; 6 : 1072-4.
79. Fromm M, Taylor LP, Walbot V. Stable transformation of maize after gene transfer by electroporation. *Nature* 1986 ; 319 : 791-3.
80. Lazzeri PA, Brettschneider R, Lührs R, Lörtz H. Stable transformation of barley *via* PEG-induced direct DNA uptake into protoplasts. *Theor Appl Genet* 1991 ; 81 : 437-44.
81. Klein TM, Arentzen R, Lewis PA, Fitzpatrick-McElligott S. Transformation of microbes, plants and animals by particle bombardment. *Bio/Technology* 1992 ; 10 : 286-91.

82. McCabe DE, Swain WF, Marinelli BJ, Christou P. Stable transformation of soybean (*Glycin max*) by particle bombardment. *Bio/Technology* 1988 ; 6 : 923-6.
83. McCabe D, Martinell BJ. Transformation of elite cotton cultivars *via* particle bombardment of meristems. *Bio/Technology* 1993 ; 11 : 596-8.
84. Fromm ME, Morrish F, Armstrong C, Williams R, Thomas J, Klein TM. Inheritance and expression of chimeric genes in the progeny of transgenic maize plants. *Bio/Technology* 1990 ; 8 : 833-9.
85. Vasil V, Srivastava V, Castillo AM, Fromm ME, Vasil IK. Rapid production of transgenic wheat plants by direct bombardment of cultured immature embryos. *Bio/Technology* 1993 ; 11 : 1553-8.
86. Becker D, Brettschneider R, Lörtz H. Fertile transgenic wheat from microprojectile bombardment of scutellar tissue. *Plant J* 1994 ; 5 : 299-307.
87. Somers DA, Rines HW, Gu W, Kaeppler HF, Bushnell WR. Fertile, transgenic oat plants. *Bio/Technology* 1992 ; 10 : 1589-94.
88. Christou P, Ford TL, Kofron M. Production of transgenic rice (*Oriza sativa L.*) plants from agronomically important indica and japonica varieties *via* electric discharge particle acceleration of exogenous DNA into immature zygotic embryos. *Bio/Technology* 1991 ; 9 : 957-62.
89. Castillo AM, Vasil V, Vasil IK. Rapid production of fertile transgenic plants of rye (*Secale cereale L.*). *Bio/Technology* 1994 ; 12 : 1366-71.
90. Ritala A, Aspegren K, Kurtén U, Salmenkallio-Marttila M, Mannonen L, Hannus R, Kauppinen V, Teeri TH, Enari TM. Fertile transgenic barley by particle bombardment of immature embryos. *Plant Mol Biol* 1994 ; 24 : 317-25.
91. Casas AM, Kononowicz AK, Zehr UB, Tomes DT, Axtell JD, Butler LG, Bressan RA, Hasegawa PM. Transgenic sorghum plants *via* microprojectile bombardment. *Proc Natl Acad Sci USA* 1993 ; 90 : 11212-6.
92. Maliga P. Towards plasmid transformation in flowering plants. *Trends Biotechnol* 1993 ; 11 : 101-7.
93. Maliga P, Carrer H, Kanevski I, Staub J, Svab Z. Plasmid engineering in land plants: a conservative genome is open to change. *Phil Trans R Soc London B* 1993 ; 342 : 203-8.
94. Joshi RL, Joshi V. Strategies for expression of foreign genes in plants, potential use of engineered viruses. *FEBS Lett* 1991 ; 281 : 1-8.
95. Chapman S, Kavanagh T, Baulcombe D. Potato virus X as a vector for gene expression in plants. *Plant J* 1992 ; 2 : 549-57.
96. Kumagai MH, Turpen TH, Weinzettl N, Della-Cioppa G, Turpen AM, Donson J, Hilf ME, Grantham GL, Dawson WO, Chow TP, Piatak M, Grill LK. Rapid, higl-level expression of biologically active α-trichosanthin in transfected plants by an RNA viral vector. *Proc Natl Acad Sci USA* 1993 ; 90 : 427-30.
97. Martineau B, Voelker TA, Sanders RA. On defining T-DNA. *Plant Cell* 1994 ; 6 : 1032-3.
98. Yoder JI, Goldsbrough AP. Transformation systems for generating marker-free transgenic plants. *Bio/Technology* 1994 ; 12 : 263-7.
99. Dale EC, Ow DW. Intra- and intermolecular site-specific recombination in plant cells mediated by bacteriophage P1 recombinase. *Gene* 1990 ; 91 : 79-85.
100. Goldsbrough AP, Lastrella CN, Yoder JI. Transposition mediated repositioning and subsequent elimination of marker genes from transgenic tomato. *Bio/Technology* 1993 ; 11 : 1286-92.
101. Ling K, Namba S, Gonsalves C, Slightom JL, Gonsalves D. Protection against detrimental effects of potyvirus infection in transgenic tobacco plants expressing the papaya ringspot virus coat protein gene. *Bio/Technology* 1991 ; 9 : 752-8.
102. Fitch MMM, Manshardt RM, Gonsalves D, Slightom JL, Sanford JC. Virus resistant papaya plants derived from tissues bombarded with the coat protein gene of papaya ringspot virus. *Bio/Technology* 1992 ; 10 : 1466-72.

103. Spiral J, Thierry C, Paillard M, Petiard V. Obtention de plantules de *Coffea canephora Pierre* (Robusta) transformées par *Agrobacterium rhizogenes*. *CR Acad Sci Paris* 1993 ; 316 : 1-6.
104. Gambley RL, Ford R, Smith GR. Microprojectile transformation of sugarcane meristems and regeneration of shoots expressing β-glucuronidase. *Plant Cell Rep* 1993 ; 12 : 343-6.
105. May GD, Afza R, Mason HS, Wiecko A, Novak FJ, Arntzen CJ. Generation of transgenic banana (*Musa acuminata*) plants *via Agrobacterium*-mediated transformation. *Bio/Technology* 1995 ; 13 : 486-92.
106. Sagi L, Panis B, Remy S, Schoffs H, De Smet K, Swennen R, Cammue BPA. Genetic transformation of banana and plantain (*Musa* spp.) *via* particle bombardment. *Bio/Technology* 1995 ; 13 : 481-5.
107. Sain SL, Oduro KK, F DB. Genetic transformation of cocoa leaf cells using *Agrobacterium tumefaciens*. *Plant Cell Tissue Organ Culture* 1994 ; 37 : 243-51.
108. Arokiaraj P, Jones H, Cheong KF, Coomber S, Charlwood BV. Gene insertion into *Hevea brasiliensis*. *Plant Cell Rep* 1994 ; 13 : 425-31.
109. Schell J. Closing lecture. ISPBM congress. *Plant Mol Biol* 1994 ; 26 : 1695-9.

6

The Regulations

Sophie Béranger*, Hervé Reverbori**

Summary

The regulations concerning the deliberate release of Genetically Modified Organisms (GMOs) are derived from directive 90/220/EEC of April 23rd 1990. Its incorporation into French law was implemented on July 13th 1992.

The regulatory system in place stipulates that any deliberate release into the environment must be authorized by the national authority.

Approval to market GMOs is granted after agreement by the relevant European Community authorities.

At the national level, for each regulated product, the requirements of directive 90/220 have been taken into account in specific "product" decrees.

For this reason, a single authorization is granted in France for the marketing of a given GMO product. By way of example, plants are used to illustrate this procedure.

* Ex-Head of the Unit for Regulatory Affairs, Ministry of Agriculture, Fisheries and Food, Paris, France.

** Head of the Unit for Regulatory Affairs, Ministry of Agriculture and Fisheries, Paris, France.

The regulations concerning the deliberate release of genetically modified organisms (GMOs) have been developed concomitantly with the increasing use of genetic engineering techniques. The overall aim of these regulations is to protect the environment and public health. One of their notable features is that they were not created to mitigate known disadvantages of genetic engineering but rather to forestall any potential risks inherent in this new technique.

The present chapter describes the basic aspects of these regulations, *i.e.*:
- directive 90/220/EEC of April 23rd 1990 concerning the deliberate release of GMOs in the environment;
- the law of July 13th 1992 and its associated regulations. The example chosen to illustrate this system involves transgenic plants due to the large number of releases of such plants.

Directive 90/220/EEC: A "Horizontal" Directive

In order to avoid competition between member states and to reply to a certain unease concerning the potential risks for humans and the environment, directive 90/220 tries to create new administrative procedures and to harmonize the existing ones for the assessment of the deliberate release of genetically modified organisms.

This directive outlines the authorization procedures for the deliberate release for research and development (part B of the directive) and the marketing consent procedures (part C of the directive).

The Field of Application

The directive applies to organisms that have been modified other than by natural multiplication or recombination.

An organism is any biological non-cellular, cellular or multicellular entity that is able to reproduce or transfer genetic material. GMOs are organisms whose genetic material has been modified by other means than natural multiplication or recombination.

The field of application does not include genetically modified organisms that are obtained by viral infection, mutagenesis, conjugation, transduction, polyploid induction, hybridoma formation and the fusion of plant cells obtained by traditional methods. For example, the fusion of protoplasts within the same botanical family is a technique that is not included in the field of application.

However, it is noteworthy that according to the directive 90/220, genetically modified organisms resulting from the use of a combination of these techniques are covered by the directive.

Authorization: Part B of Directive 90/220

The directive indicates that an application in compliance with the requirements of Appendix II should be sent to the competent national authorities for assessment and consent.

A summary file is sent to the other European member states for information.

Appendix II of the directive was recently specifically adapted for plants, simplifying certain points. As a result, the summary file was also modified and simplified.

The Procedures for Marketing Consent (Part C of the Directive)

The term "marketing consent" used in directive 90/220 is a source of ambiguities when used within the context of the French regulations.

At the national level, marketing consent is often interpreted as a sales authorization.

Within the content of directive 90/220, the consent involves the evaluation of the risks for humans and the environment related to the commercialization of a product consisting of GMOs. For this reason, it is preferable to speak of consent according to part C of the directive.

This consent procedure, is under the authority of the European Commission, and complies with the requirement for free circulation of these new products within the European market.

The different steps in the procedure involve:
• the party that wants to release a GMO for the purpose of commercialization submits a technical application to the competent authorities of one of the member states of the Community (chosen by the applicant). The application should include the pertinent information indicated in Appendices II and III of the directive;
• the competent authority evaluates the risks to humans and the environment and gives or withholds consent within 90 days from the date of the acknowledgement;
• in case of a favourable decision, the file is sent from the refereeing authority to the European Commission that immediately sends it to the competent authorities in the other member states;
• the release is authorized by the competent authority within 60 days if no objections are raised by the other member states;
• if objections are raised, the release may be authorized only if a qualified majority vote by the Committee handling the deliberate release of GMOs in the environment should authorize the release. The committee consists of representatives of the member states and is chaired by a Commission representative.

A qualified majority decision is obtained with at least 62 out of 87 votes. The votes are weighted as follows: Germany, France, Italy and the United Kingdom, 10 votes; Spain, 8 votes; Belgium, Greece, Netherlands and Portugal, 5 votes; Austria and Sweden, 4 votes; Denmark, Ireland and Finland, 3 votes; Luxembourg, 2 votes.

If a qualified majority decision is not obtained, the decision may be modified and adopted by the Council of Ministers of the Environment.

The Council has three months to decide.

If the Council does not decide within three months, the Commission's decision is adopted.

Directive 90/220 provides for alterations in this procedure when the European regulations concerning a product (a plant for example) takes the requirements in this directive into account. The specific procedure concerning the product then applies.

The Law of July 13th 1992 and the Regulatory System

A Wide Field of Application

Conceived to apply the European directives 90/219 concerning the restricted use of genetically modified micro-organisms and 90/220, the law of July 13th 1992 covers a wide area. Whereas directive 90/219 solely covers the restricted use of genetically modified micro-organisms, the French parliament has deliberately chosen to include the use of both micro-organisms and other GMOs.

This widening of the range covered by the regulations is due to the technical difficulties in defining a clear-cut separation between micro-organisms and organisms, and also to maintain a coherent approach to the evaluation of risks linked to the use of GMOs.

However, not all GMOs fall under the law. Certain so-called traditional techniques (a list of these techniques is indicated in decree 93-774 of March 27th 1993) may be used to obtain GMOs without statutory restrictions.

Special attention is used in defining an organism. A non-living GMO that is able to transfer genetic material, such as for example a killed bacteria whose plasmids are able to transform other bacteria, is covered by the law. The same reasoning applies to naked DNA.

Transport of GMOs is not covered by this law but falls within the jurisdiction relative to the transport of dangerous substances. GMOs fall within category 6.2

(infectious substances) or 9 (sundry products that are dangerous for the environment).

The law distinguishes between the provisions concerning restricted use and those concerning the deliberate release and commercialization of GMOs. A GMO can only be possessed or *a fortiori* used within the framework of one of these regulations. We will only deal with the deliberate release and commercialization of GMOs.

Deliberate Release

Deliberate release is defined as the deliberate introduction of a GMO into the environment for research and development purposes or any other purpose than marketing.

Deliberate release requires consent granted by the administrative authority in charge of marketing regulations for the product under consideration. For example, deliberate release of a GMO drug is granted by the Agence du Médicament while the release of a transgenic plant is granted by the Ministry of Agriculture. This involves different sector-specific decrees, which French law, by applying one of the guiding principles of directive 90/220, has anticipated. In all cases the Biomolecular Engineering Commission (for details, refer to the chapter on the BEC) is consulted about the deliberate release of every product derived from genetic engineering. This provides horizontal consistency in the processing of the different applications.

The law establishes the principle of the public's right to information concerning the effects on public health and the environment linked to the release of a GMO and exactly defines the information that cannot be considered as confidential.

Commercialization

Commercialization is defined as making products consisting partly or wholly of GMOs available to third parties, with or without cost. This wide-ranging definition does not assume that there needs to be a commercial transaction for a product to be put on the market. Marketing consent is granted by a national administrative authority after tacit or explicit approval by the community authorities. It is valid for all countries in the European Union. If the product to be commercialized is covered by a specific application for authorization and licensing, independent of the fact that it consists of a GMO, a single authorization is granted *via* this specific procedure under the law of July 13th 1992. As with deliberate release, the product under consideration falls within the scope of the responsible administrative authority.

Obligations and Sanctions defined by the Law

Since this technology is new, with potential effects which are sometimes difficult to assess, the law provides a number of guarantees together with legal or administrative sanctions, if required.

The administration should be informed of any new element that may influence the assessment of the risks for public health or the environment. The holder of an authorization under the law of July 13th 1992 is still responsible even after providing the information since he must, by his own initiative, take all of the steps required to protect the environment and public health.

Under this law, the administrative authority has a great deal of power. At any time, it can suspend or withdraw authorization, impose changes in the conditions for release and order the destruction of GMOs or ban the sale of products. Except in an emergency, these steps are only taken once the holder of the authorization has been able to comment.

In case of non-compliance with the regulations, the administrative authority may, after formal notice, order the offending products to be consigned.

Finally, the violation of the different provisions in the law will be penalized by a range of legal sanctions: a fine of 10,000 to one million francs and/or a prison sentence of two months to two years depending on the type and gravity of the act.

For each of the products where specific regulations exist, the requirements of directive 90/220 have been taken into account in the "product" decrees. The list of these decrees is provided in *Table I.*

A procedure is examined by way of example afterwards.

The Case of Transgenic Plants

Decree 93-1177 of October 18th 1993 establishes the conditions that apply to the deliberate release in the environment and the marketing of genetically modified plants.

The order of September 21st 1994 establishes the elements required in applications for the deliberate release and marketing of genetically modified plants or seeds.

The consent for research, development and marketing are granted by the Minister of Agriculture after agreement by the Minister of the Environment. Public information is provided by the local councils where the release is carried out.

Table I. List of "Product" Texts.

Name of the Text	Date of Publication
1 - Decree 93-1177 of October 18th 1993 for applications concerning plants and seeds listed in category III in the Law of 13th July 1992.	Official Gazette of October 20th 1993
2 - Decree 94-510 of June 23rd 1994 concerning the marketing of ornamental plants, young vegetable plants, fruit plants and the propagation material of all of these plants.	Official Gazette of June 24th 1994
3 - Order of September 21st 1994 concerning requests for deliberate release into the environment for any purpose other than commercialization and concerning marketing consent applications for plants, seeds or genetically modified plants.	Official Gazette of October 18th 1994
4 - Decree 94-46 of January 5th 1994 establishing the conditions for the deliberate release of genetically modified organisms: (i) destined for human food (other than plants, seeds, or animals), (ii) used in the composition of products for cleaning materials and objects coming in contact with foods, products or drinks for human or animal consumption.	Official Gazette of January 19th 1994
5 - Decree 94-359 of May 5th 1994 concerning the control of phyto-pharmaceutical products.	Official Gazette of May 7th 1994
6 - Decree 95-487 of April 28th 1995 establishing the conditions for the deliberate release of genetically modified animals.	Official Gazette of April 30th 1995
7 - Decree 95-1172 of November 6th 1995 establishing the specific procedures for genetically modified organisms intended to be used in drugs for humans and the products mentioned in 8, 9 and 10 of article L.511-1 of the Public Health Code.	Official Gazette of November 8th 1995
8 - Decree 95-1173 of November 6th 1995 establishing the specific procedures applicable to genetically modified organisms to be used in drugs for veterinary purposes.	Official Gazette of November 8th 1995
9 - Decree 96-317 of April 10th 1996 concerning the control of the use of genetically modified organisms, as regards genetically modified elements or products derived from human tissues.	Official Gazette of April 13th 1996
10 - Decree 96-850 of September 20th 1996 concerning the control of the deliberate release and commercialization for civil purposes of products in whole or in part made of genetically modified organisms.	Official Gazette of September 27th 1996
11 - Decree 97-685 of May 30th 1997 concerning the control of the use and the deliberate release of genetically modified organisms for feed purposes.	Official Gazette of June 1st 1997
12 - Decree 98-318 of April 28th 1998 concerning the control of fertilising materials and culture support composed as whole or part of genetically modified organisms	Official Gazette of April 29th 1998

As regards commercialization, the French system includes authorization (part C) of directive 90/220 in the existing marketing consent procedure (registration in the catalogue).

For a genetically modified variety, the Minister of Agriculture refers the matter to two committees: the Comité Technique Permanent de la Sélection des Plantes Cultivées (CTPS) and the Biomolecular Engineering Commission (in French: CGB).

For all varieties that are candidates for registration in the catalogue, the CTPS verifies whether or not they are genetically modified, the difference, homogeneity and stability of these varieties as well as, for most of them, their agricultural and technical value.

The BEC specifically assesses genetically modified organisms as a function of their potential risk for humans and the environment (including the assessment required for part C of directive 90/220/EEC).

It delivers an opinion to the Minister of Agriculture within sixty days of receiving an application. If the opinion is favourable, the Minister of Agriculture sends the application, within ninety days of receipt, to the European Commission. The Community procedure then follows the conditions described above.

If the community decision is favourable, the Minister of Agriculture asks the CTPS to conclude the procedure.

This procedure for genetically modified plants therefore initially allows for the parallel consultation of the CTPS and CGB.

However, it may be adapted to suit specific cases. For example, an industrialist may wish to obtain authorization corresponding to part C of directive 90/220 for plants that have reliably integrated the desirable character, but the varieties marketed and registered in the catalogue will be derived from these plants by cross-fertilization using traditional methods.

In this case, the industrialist can file a part C application with the Minister of Agriculture who then only refers the matter to the CGB.

If this committee approves, the minister forwards the file to the European Commission as described above.

New development on GMO regulations

Novel food and novel ingredient regulation n°258/97 entered into force on May 15th 1997. It scope covers foods (including ingredients) which have not hitherto been used for human consumption to a significant degree within the European

Community. Namely those foods are especially food containing GMOs, food products from those GMOs. To obtain authorisation or to be legally marketed, they must comply with the following principles :
- not to present danger to consumer,
- not to mislead the consumer
- not to induce nutritionally disadvantageous for the consumer.

Products are subject to a single safety assessment through a community procedure before they are placed on the market.

Additional specific requirement on labelling are compulsory in order to assure that the necessary information is available to the consumer (ethical concerns, presence of GMOs, new characteristics...).

Regulation n°1139/98 provides complementary specific labelling provisions for products and ingredients derived from soya and maize genetically modified. Since DNA fragments or proteins from genetic event are present, compulsory mention on labelling for final consumer must indicate that the ingredient is produced from GMOs.

The European Commission has adopted the principles of a proposal for modification of the directive 90/220/EEC on the deliberate release into the environment of genetically modified organisms. This proposal amends the existing directive. This proposal improves provisions for labelling, introduces the systematic consultation of European scientific committees, introduces mandatory monitoring of products after their placing on the market which will be linked to a consent granted for a fixed period, increases the transparency of the decision making process and modifies the comitology provisions, confirms the possibility to raise ethical concerns and clarifies the scope of the directive. The proposal sets out common principles for risk assessment under the directive.

Conclusion

The statutory flexibility of the French system is exemplary in the European Union and should be maintained as far as possible. It has harmoniously accompanied major research efforts carried out in the plant biotechnologies since France is the leading country in the Union in terms of the number of deliberate releases of transgenic plants.

This system, based on specific procedures for GMOs in each product sector, has been able to find the right balance between the scientific and economic imperatives and a high level of protection of public health and the environment. It is based on the following concept: although, as a new technology, genetic engineering requires specific assessment, as soon as the lack of risk for humans and the environment has been established, there is no reason to discriminate statutorily between a GMO product and a product obtained using traditional technologies.

Unfortunately, this idea is not shared by all countries in the European Union, largely due to the pressure from public opinion, which is unfavourable towards genetic engineering in Northern Europe. Decisions of major economic importance (marketing consent) are taken at the community level. Such a situation reinforces the role of the Commission which, in addition to its important statutory powers, has the difficult task of mediation.

The recent positions taken by the Commission and the Council tend towards the development of biotechnology. The pertinence of the current directives was reviewed, in order to make sure that the procedures do not hinder the development of European biotechnologies in the face of world competition.

In the farming sector, for example, the United States will certainly not fail to use all possible means to reinforce their competitiveness, including biotechnology, that they are likely to make one of the most important factors for future economic growth.

7

Gene Flow

Pierre Thuriaux*

Summary

*Genetic traits which have a selective advantage in agricultural or natural systems are liable to spread beyond the cultivated variety in which they were originally introduced. This applies to herbicide resistance, but also to characters such as increased tolerance to pests (insects, fungi, bacteria or viruses) or to environmental stress (drought, frost, etc.). A given character (which may or may not be transgenic) could be transferred to other varieties of the same crop by cross-fertilization (intraspecific gene flow), which may, for example, generate varieties that are resistant to several herbicides. Less frequently, it may also be transferred to related plant species having some interfertility with the original variety (interspecific gene flow). How far this may affect a given trait in a given crop will very much depend on the physiological and genetic properties of the plant species, on the agricultural context and on the nature of the transgenic character itself. Rape (*Brassica napus*), beet (*Beta *sp.) and sunflower (*Helianthus *sp.) are the main crops where gene flow is likely to occur.*

*Unlike most characters obtained by empirical selection, transgenic traits are dominant and monogenic, and are thus readily transmittable by genetic outcrossing. Moreover, many of these characters simply cannot occur without transformation. This is obvious in the case of insect resistance mediated by the bacterial *Bt *toxin, but also applies to resistance against*

* Ex-member of the CGB, Senior Scientist, Commissariat à l'Énergie Atomique, Saclay, France.

such herbicides as glyphosate or glufosinate, for which few or no cases of spontaneous resistance have been noted to date. The release of dominant resistant alleles created ex nihilo *in transgenic plants therefore creates a novel situation in terms of gene dissemination. Moreover, using herbicide resistance as a genetic marker to monitor the inheritance of other transgenic characters is an effective way of ensuring the* de facto *dissemination of such resistance. This practice should be used with the greatest caution.*

There is no evidence that intra- or interspecific dissemination has ever taken place in controlled field releases conducted so far in France or elsewhere, but this merely shows that it does not happen on a massive scale. Large-scale experiments (especially in plants such as rape and beet) are needed to monitor extra-varietal dissemination in real-life situations, in order to have a realistic view of the problem and to adjust risk assessment policies if necessary. Generally speaking, extra-varietal dissemination creates some uncertainty as to the long-term ecological impact of transgenic plants. This is one of the reasons why transgenic constructs should be restricted to minimal insertions of foreign genetic material, i.e. the gene(s) of interest and a suitable selection marker. The genetic structure in planta *should be carefully checked by molecular analysis. By the same token, constructs combining a universal bacterial origin of replication* (e.g. *the oriV element) and an antibiotic resistance marker must be avoided, in view of the remote possibility of gene transfer to bacterial organisms.*

Factors Affecting Intervarietal and Interspecific Gene Flow

In a given population, each gene (whether transgenic or not) is represented by different allelic forms that have a certain probability of becoming established in the population or in the whole species, with a tendency towards equilibrium if exchanges are rapid and if the environment remains relatively constant. As discussed below, the intensity of gene flow primarily depends on the allogamy of the species, on the genetic or geographic isolation of the variety and on the positive or negative selection pressure exerted on the character under consideration. The likelihood of gene flow towards another variety of the same plant, or even another species, must therefore be estimated on a case by case basis. Depending on the transgenic character considered, one must then assess the negative impact that may be associated with gene flow effects on consumers, users and producers of the transgenic plant.

Biologists draw a clear-cut distinction between the notions of species, population and variety. A species includes all the interfertile individuals able to exchange their genes by cross-fertilization, whereas a population is made of all the individuals forming a distinct and homogenous group as far as certain genetic characters are concerned. The concept of variety, often used in plant genetics, refers to populations separated by an especially pronounced reproductive isolation. Domestic varieties are an extreme example. This useful but vague concept is an endless source of controversy between specialists. The main point is that a species is defined by the interfertility of its individual members and by the strong barriers to the transfer of genes outside the species.

Hence, there is a major difference between intervarietal gene flow (which occurs frequently, at least in allogamous plants) and interspecific gene flow, a much rarer event, that only occurs between closely related species. In addition, it has been argued that transgenic genes (in particular those carrying antibiotic resistance) may even be transferred to bacterial genomes. Contrary to widespread belief, this is rather an implausible scenario, which needs however to be carefully assessed in view of current fears that transgenic plants may contribute to the unfortunately very real spread of antibiotic resistance in bacterial populations.

Intervarietal gene flow is by no means limited to transgenic traits, but transgenic plants are to some extent a special case. First, transgenic characters (at least those currently approaching commercialization, such as *Bt* toxin production, herbicide-resistant plants, etc.) correspond to the *ex nihilo* acquisition of one or several foreign genes. In most cases, no equivalent character would have appeared in the plant by spontaneous or induced mutation or by cross-breeding. Second, they are monogenic and dominant, and thus readily expressed upon dissemination in another genetic context. Third, some of the transgenic characters currently developed (herbicide tolerance, resistance to insects or viruses) are bound to be advantageous in some agricultural environments (if the herbicide is used on other crops) or even in the natural environment (resistance to insects or viruses). In the case of herbicide-tolerance, an additional problem is the foreseeable emergence of multiresistant plants by intervarietal cross-fertilization.

A key factor in gene dissemination is obviously the fitness of the transgenic character in a given ecosystem. The ecological performance of control rape and transgenic plants that are resistant to kanamycin or the herbicide Basta (glufosinate) has been examined in several types of habitat [1]. These plants did not have a distinct competitive advantage or disadvantage. However, this study merely shows that the transgenic plant is not especially vigorous without herbicide treatment, and it is not possible to conclude as to its long-term dissemination, especially in an environment where the use of Basta is encouraged. Genes with a high selective cost are bound to disappear rapidly from the population in the absence of specific selection pressure but, contrary to a widespread idea, need not have a high selective value in order to persist, considering the complexity of the selection factors or even the partially neutral character of evolution. In addition, the position of the transgene on the chromosome may favour its dissemination if it is genetically linked to a character with a high selective value ("hitch-hiking effect"). On the whole, it is useful to distinguish between transgenic characters without any manifest selective advantage or even a probable disadvantage (modification of the food or gustatory properties, genes facilitating processing, etc.) and those with at least a probable selective advantage in certain agricultural contexts (herbicide resistance) or even in the natural ecosystem (resistance to insects or other phytopathogens).

Gene flow also greatly depends on the reproductive physiology of the plant involved. Several aspects are determinant: the probability of cross-pollination, the vigour of hybrid plants and of their seeds, the ability of their seeds to germinate after a long dormancy, and the fertility of the F1 hybrids. Cross-pollination mainly depends on the type of pollination (by wind or insects) the type of farming (rotation, fallow) and the possibility of regrowth in the following years. In addition to genetic and physiological parameters directly related to the plant itself, local geography and farming practices may dramatically affect the likelihood of intervarietal gene transfer. Interspecific gene transfer to related species is much rarer than intraspecific exchanges. To begin with, many European crops originate from the New World and have no wild relatives in Europe. Second, and most importantly, interspecific exchange is usually prevented by fertility barriers that exist even between closely related species (after all, the Darwinian definition of two distinct species is that they cannot cross-fertilize). As discussed below, crops such as rape, beets and sunflowers are nevertheless likely to participate in a low-level of interspecific gene flow in natural conditions.

Impact of Gene Transfer Effects in some European Crops

Most European crops are, in effect, largely isolated from interfertile varieties and from related wild species. Tobacco, potato and maize, for instance, originate from South America and have simply no indigenous relatives in Europe (leaving aside the interfertility of tobacco with some ornamental tobaccos). Maize is so domesticated in Europe that the plant would not survive without man and it is highly unlikely that the seeds obtained from intervarietal cross-fertilization (in itself probable due to the high allogamy of maize) would provide descendants. Allogamy is rare (although possible) in the case of potato and tobacco and the dissemination of their seeds is highly unlikely due to the physiology of these two plants. Other plants, originating in the Old World, such as wheat (for which transgenic varieties are still far from the market, due to the very low transformation efficiency obtained with this plant) and soybean are genetically isolated by the allopolyploid structure of the wheat genome, by their low cross-fertilization and by the poor persistence of seeds. In all these plants, the effectiveness of intervarietal gene exchange is deemed to be fairly limited or even negligible. In particular, the formation of herbicide-multiresistance by cross-fertilization between cultivated varieties will probably be very slow in the

European context. However, it is necessary to pay attention to the increasing internationalization of the seed trade in this respect.

In plants such as rape [2-4], sunflower, chicory [5] (the latter being a special case, considering its special type of cultivation) or beet [6,7], gene flow is likely to be a recurrent phenomenon. Rape has a substantial rate of allogamy and the seeds are highly persistent. The appearance of herbicide-multiresistant plants will almost certainly follow the commercialization of resistant plants. Sunflower (*Helianthus* sp.) should essentially behave like rape in this respect. The case of sugar or fodder beet (*Beta* sp.) is somewhat different since this biennial plant is systematically harvested before flowering which in principle, eliminates cross-fertilization and thereby the appearance of multiresistance. However, sugar beet fields can be highly contaminated by wild, "weedy" beets, and the emergence of herbicide-multiresistant beets is, in the long-term, a likely event. Thus, there is a distinct possibility that herbicide-resistance genes will spread in rape, sunflower and beets, eventually leading to multiresistant plants. Moreover, the use of herbicide-resistance as a genetic marker (linked to other genes of interest) may considerably favour its dissemination. Whether or not this will be a fast and general process is, at present, very difficult to guess. In the future, herbicide-resistant rape or wild beet may to some extent be the "weeds" of other crops, which is not the case at present. Many farmers and agronomists argue that, in practice, this would not be a major issue. Others, instead, are seriously concerned. Be it as it may, the commercialization of such varieties would certainly need to be accompanied by an effective monitoring system allowing national and European authorities to detect any serious agronomic problem related to their dissemination and, if needed, to withdraw them from the seed market.

Interspecific gene flow is much less frequent than intervarietal exchanges. Again, rape and beets are quite distinct. Rape is a natural hybrid between a cabbage (*Brassica oleracea*) and a turnip (*Brassica campestris*) and is interfertile with the latter [2,8]. Moreover, it has a low degree of interfertility with several "weedy" species such as *Raphanus raphanistrum* or much more rarely, hoary mustard (*Sinapis arvensis*) [9,10]. This produces hybrid plants which, after further generations, carry the herbicide resistance gene and are genetically close to the wild relative (in fact, they have acquired one or a few additional chromosomes originating from rape, including the chromosome that bears the herbicide-resistance gene). Whether these hybrids will be both frequent and vigorous enough to create a serious problem in rape cultures (or even other crops) is an entirely open question. Cultivated beets, on the other hand, are known to cross-fertilize "wild" beets that flower every year (whereas cultivated beets are biennials and are harvested before flowering). Initially, it will obviously be much easier to eliminate these "wild" beets, which are a major plague for farmers, if the crop beets are herbicide resistant. In the long run, however, one can anticipate that "wild" beets (including distinct but closely related species such as *B. maritima* and *B. campestris*) are likely to eventually acquire the herbicide-resistance gene [6,7].

This of course, will mean that beet producers will be back to the current situation where "wild" beets, turning to seed early, are eliminated by hand from seed preparation fields. As far as other cultures are concerned, wild beets are currently not a problem, but the situation may change if they acquire herbicide resistant genes.

Exchanges between Plants and Bacteria: Can Antibiotic-Resistance Genes be Transferred back to Bacteria?

Under rather special circumstances, bacteria can take up foreign DNA by genetic transformation (or phage transduction) and can integrate this DNA into their own genome by genetic recombination. However, in the field, this would depend on three conditions: the persistence in the soil (or the digestive tract) of sufficiently long fragments of DNA to contain the whole transgene, the presence of spontaneously transformable bacteria and the overcoming of the barriers preventing the integration of exogenous DNA, due to its lack of sequence homology with the bacterial genome [11]. Based on current knowledge, such exchanges are highly unlikely and attempts to demonstrate them have been unsuccessful. A second mode of bacterial gene transfer relies on plasmid-mediated conjugation (plasmids are small genomic elements that reproduce themselves within the bacterial cells). Plasmids have a remarkable genetic fluidity due to the action of transposable elements. In recent times, this ability has been revealed by the dissemination of antibiotic resistance genes among a variety of bacterial species, a very spectacular case indeed of gene flow amplified by human intervention [12]. Some conjugative plasmids have a remarkably broad host spectrum, and can also mobilize non-conjugative plasmids that are themselves able to reproduce in a large variety of bacteria, due to the presence of a wide host-range replication origin (oriV). Such exchanges were observed in the laboratory between bacterial species that are taxonomically very distant, such as gram positive and gram negative bacteria (such as *Escherichia coli*) and cyanobacteria [13,14]. Their existence has been demonstrated in soils [15]. Other biotopes (silos, the rumen of ruminants) contain huge microbial communities where such exchanges may occur. By extension, the natural transfer of genetic material from bacteria to higher organisms (plants, fungi, protozoans, nematodes) is known to be possible under certain circumstances. Conjugative plasmid transfer between *E. coli* and yeast can be carried out in the laboratory [16], but bacterial plasmids cannot stably establish themselves in yeasts, since naturally existing yeast plasmids are entirely unrelated to the bacterial ones. The *Agrobacterium* plasmid Ti is so far the only example of natural – albeit very specialized – system of gene transfer between bacteria and plants, resulting in the integration of a bacterial DNA fragment (T-DNA) into the genome of the infested dicot, followed by crown-gall formation. As discussed elsewhere in this book (see the chapter by F. Casse), this system has been put to good use as vector for plant transformation. The transfer of T-DNA was for a long time thought to occur only in dicots, but recent results indicate a much wider transfer spectrum, since T-DNA can be transferred to the genome of monocots such as rice [17], and even to yeast [17,18]. Moreover, the similarity between the Ti plasmid and bacterial conjugative plasmids is striking [19] . Gene transfer systems operating from bacteria to plants or higher organisms may therefore be more frequent than is generally believed, and could indeed play some role in evolution [20].

Can gene transfer occur in the opposite direction, *i.e.* from transgenic plants to bacteria? This is still mere speculation, but not one that can be ignored in terms of risk assessment. It should be kept in mind that the genes for herbicide, antibiotic and even insect resistance (in the case of the *Bt* toxin) currently introduced into transgenic plants all originate from soil bacteria. The eventuality of their return to the soil microbial flora raises therefore no special concern. The possibility that an antibiotic resistance gene present in a transgenic plant could be transferred back to bacteria, therefore spreading new resistance to antibiotics, is sometimes mentioned as a (major) hazard of transgenic plants. Antibiotic resistances are, of course, a serious problem in public health, due to the careless over-prescription of antibiotics and, perhaps even more importantly, to their widespread use in animal farming [21]. It is very hard to see how the transfer of an antibiotic resistance gene from a plant to an animal or human bacterium (which, as mentioned above, is a very unlikely event that has never been observed so far) could contribute in any significant way to the present situation. The genes for resistance to the antibiotics (kanamycin, ampicillin) used in the selection of transgenic plants are, unfortunately, already quite widespread in human bacteria. Nevertheless, the genetic context in which they are introduced in transgenic plants must be carefully considered. In particular, it is certainly unwise to insert them along with a bacterial origin of replication, especially a universal one like the oriV element, that may, under certain circumstances, be introduced into transgenic plants generated by *Agrobacterium* transformation. In addition, the possibility of plasmid exchange between different bacteria must be very strictly taken into account when using transgenic bacteria in industrial processes.

Conclusion

Any transgene has a non-negligible probability of leaving its initial variety to spread in interfertile varieties of the same species as long as they coexist geographically. In practice, and within the European context, this possibility is high for herbicide-resistance genes (or other traits with a distinct selective advantage) in species such as rape, beet and sunflower. The possibility of introgression in other related species is low but real for certain plants such as rape or beet. The main problems associated with such intervarietal or even inter-specific gene flow will be related to the emergence of herbicide multiresistant crops which, in some cases, may notably reduce the usefulness of the corresponding herbicides. The transfer of transgenic characters towards the surrounding bacteria flora is very unlikely, even if the reality of the phenomenon on the scale of evolution is scientifically plausible.

These considerations underline the need for careful monitoring of gene flows even if, in most cases, the anticipated ecological consequences are negligible and do not radically question the agronomic, industrial or medical use of genetically modified organisms. Humility is required in this area. The accelerated dissemination of antibiotic resistance characters in human (and farm animals) or, in another context, the sensitivity to *Helminthosporium* associated with the cytoplasmic male sterility of maize [22] illustrate the unpredictable effects of gene flow (non-transgenic in these two cases) provoked by technological innovations.

What can be done in practice? First, admit our limited knowledge regarding gene flow in plants, and develop basic knowledge in this hitherto poorly studied domain. Second, clearly distinguish between transgenic characters that lack any selective advantage (modification of the food or gustatory properties, genes facilitating industrial processing, etc.) from those with a probable selective advantage in agricultural contexts (herbicide resistance) or even in natural ecosystems (resistance to insects or phytopathogens). The genetic construction itself should not contain unknown or undefined genetic elements, or unnecessary genes like markers that are of no use for selection, or a plasmid origin of replication, especially those with a wide host range. The structure of the transgenic DNA needs to be carefully checked *in planta* by standard molecular techniques. In the case of herbicide-resistant varieties (or, in another context, of insect-resistant plants), effective monitoring systems must accompany their dissemination in order to detect any substantial leak of the transgene outside of the original variety and, if necessary, to withdraw these varieties from the seed market.

References

1. Crawley MJ, Hails RS, Rees M, Kohn D, Buxton J. Ecology of transgenic oilseed rape in natural habitats. *Nature* 1993 ; 363 : 620-3.
2. Jôrgensen RB, Andersen B. Spontaneous hybrisization between oilseed rape (*Brassica napus*) and weedy *B. campestris* (*Brassicacea*): a risk for growing genetically modified oilseed rape. *Am J Bot* 1994 ; 81 : 1620-6.
3. Kerlan MC, Chèvre AM, Eber F, Baranger A, Renard M. Risk assessment of outcrossing of transgenic rapeseed to related species: interspecific hybrid production under optimal conditions with emphasis on pollination and fertilization. *Euphytica* 1992 ; 62 : 145-53.
4. Scheffler JA, Parkinson R, Dale PJ. Frequency and distance of pollen dispersal from transgenic oilseed rape (*Brassica napus*). *Transg Res* 1993 ; 2 : 356-64.
5. Lavigne C, Manac'h H, Guyard C, Gasquez J . The cost of herbicide resistance in white-chicory: ecological implication for its commercial release. *Theor Appl Genet* 1995 ; 91 : 1301-8.

6. Boudry P, Saumitou-Laprade P, Vernet P, Van Dijk H. The origin and evolution of weed beets: consequences for the breeding and the release of herbicide-resistant transgenic sugar beets. *Theor Appl Genet* 1993 ; 87 : 471-8.
7. Santoni S, Bervillé A. Evidence for gene exchange between sugar beets (*Beta vulgaris L.*) and wild beets: consequences for transgenic sugar beets. *Plant Mol Biol* 1992 ; 20 : 578-80.
8. Mikkelsen TR, Andersen RB, Jôrgensen R. The risk of crop transgene spread. *Nature* 1996 ; 380 : 31.
9. Chèvre AM, Eber F, Baranger A, Renard M. Genetic mechanisms of intergeneric gene flow from transgenic crops. *Nature* 1997 ; 389 : 924.
10. Lefol E, Danielou V, Darmency H, Boucher F, Maillet J, Renard M. Gene dispersal from transgenic crops: growth of interspecific hybrids between oilseed rape and the wild hoary mustard. *J Appl Ecol* 1995 ; 32 : 803-8.
11. Rayssiguier C, Thaler DS, Radman M. The barrier to recombination between *E. coli* and *S. typhimurium* is disrupted in mismatch repair mutants. *Nature* 1989 ; 342 : 396-402.
12. Davies J. Vicious circles: looking back on resistance plasmids. *Genetics* 1995 ; 139 : 1465-8.
13. Kreps S, Férino F, Mosrin C, Gerits J, Mergeay M, Thuriaux P. Conjugative transfer and autonomous replication of a promiscuous IncQ plasmid in the cyanobacterium *Synechocystis* PCC 6803. *Mol Gen Genet* 1990 ; 221 : 129-33.
14. Schöfer A, Kalinowski J, Simon R, Seep-Feldhaus AH, Pÿhler A. High-frequency conjugal plasmid transfer from gram-negative *Escherichia coli* to various gram-positive coryneform bacteria. *J Bacteriol* 1990 ; 172 : 1663-6.
15. Top E, De Smet I, Verstraete W, Dijkmans R, Mergeay M. Exogenous isolation of mobilizing plasmids from polluted soils and sludges. *Appl Env Microb* 1994 ; 60 : 831-9.
16. Heinemann JA, Sprague G. Bacterial conjugative plasmids mobilise DNA between bacteria and yeast. *Nature* 1989 ; 340 : 205-9.
17. Hiei Y, Ohta S, Komari T, Kumashiro T. Efficient transformation of rice (*Oryza sativa L.*) mediated by *Agrobacterium tumefaciens* and sequence analysis of the boundaries of the T-DNA. *Plant J* 1994 ; 6 : 271-82.
18. Bundock P, Den Dulk-Ras A, Beijersbergen A, Hooykaas PJ. Trans-kingdom T-DNA transfer from *Agrobacterium tumefaciens* to *Saccharomyces cerevisiae*. *EMBO J* 1995 ; 14 : 3206-14.
19. Kado CI. Promiscuous DNA transfer system of *Agrobacterium tumefaciens*: role of the virB operon in sex pilus assembly and synthesis. *Mol Microbiol* 1994 ; 12 : 17-22.
20. Bork P, Doolittle PF. Proposed acquisition of an animal protein domain by bacteria. *Proc Natl Acad Sci USA* 1992 ; 89 : 8990-4.
21. Davies J. Bacteria on the rampage. *Nature* 1996 ; 383 : 219-20.
22. Levings CS, Siedow JN. Molecular basis of disease susceptibility in the Texas cytoplasm of maize. *Plant Mol Biol* 1992 ; 135-47.

8

Herbicide-Resistant Transgenic Plants

Michel Aigle*, Yves Chupeau,**
Eric Schoonejans***

Summary

The number of applications for the development, use and commercialization of transgenic plants submitted to the French Biomolecular Engineering Commission is constantly on the increase. Today, this problem is especially important due to the imminent commercialization of these plants and therefore the very large surfaces involved. The issues that need to be debated are:
1) the possible improvement of the economic and technical efficiency of weeding and the use of environmentally friendly products via *the novel possibilities provided by transgenic plants;*
2) a careful examination of the risks associated with the eventual dissemination of the herbicide resistance character outside the cultivated transgenic variety or due to modifications in herbicide use and management after marketing such varieties.
The evaluation of the prospects and risks inherent in the marketing of herbicide-resistant transgenic plants depends on the plant species, the nature of the herbicide and the introduced resistance gene, and on the

* Ex-Vice-President of the CGB, Professor, IBGC/Université de Bordeaux, Bordeaux, France.
** Member of the CGB, Senior Scientist, INRA, Versailles, France.
*** Unit for Regulatory Affairs, Ministry of Agriculture and Fisheries, Paris, France.

farming practices associated with the combined use of transgenic plants and their corresponding herbicide. In particular, this chapter reviews the current technical possibilities, which depend on the mechanisms of the resistance for the different herbicides for which resistant transgenic plants are ready to be commercialized or are being tested under field conditions.

The need to control weeds in agriculture and the unavoidable use of herbicides as opposed to mechanical weeding has led to farming practices associating compatible herbicide/plant pairs.

Although they are sources of potential difficulties when they arise naturally in adventitious plant populations due to poorly thought out farming practices, herbicide resistance mutations are powerful tools for understanding and managing the use of herbicides. Mutagenesis, followed by the selection of mutants by biochemical screens [1], also contributes to the characterization of the resistance mechanisms [2].

For the last few years, the study of resistance mechanisms (physiology, genetics, molecular basis, population genetics) associated with the data concerning the development and impact of products (stability, mobility and toxicity) provide invaluable information for the rational use of specific products. Finally, the molecular characterization of the basis of the resistance now enables the cloning of the gene involved and its transfer to plants of agronomic interest using genetic engineering techniques. These resistance genes are either obtained from plants or other organisms. Soil bacteria are the major natural sources of genes of interest.

During this period of conceptual and methodological elaboration, one of the pilot directions involves the research, characterization and then transfer of non-selective or total herbicide resistance genes to cultivated plants. Certain resistance genes have been transferred to many species (maize, rape, tomato, beet, poplar) and their function has been tested in field conditions.

The large-scale use of these herbicide-resistant transgenic plants may improve the economic and technical efficiency of weeding in crops where there are still weed problems. It may also favour the use of more environmentally friendly herbicides.

These advantages will not be maintained if populations of herbicide-resistant weeds develop in the treated zones. These populations may have two origins that are directly related to the resistant crop. One origin is the regrowth of resistant transgenic plants in subsequent crops (the case of rape), the second is the progeny of hybrids between resistant plants and related wild plants when they are interfertile (the case of beet). These populations may also develop by the selection of spontaneous mutations under the effect of strong selection pressure due to repeated herbicide treatments with the same active ingredient.

Three major groups of herbicides are currently considered in different contexts. One group acts on amino acid biosynthesis, a second group on photosynthesis and a third group whose action has still not been fully assessed.

The vast majority of projects in progress involve glyphosate or phosphinothricin. Product quality and the experience accumulated during their use also make them plausible candidates for European approval. Since there is no spontaneous resistance to these herbicides, the essential aspect of the problem involves population genetics. A novel character is introduced, coupled (or not) with selection pressure in its favour [3]. This approach will transform these non-selective herbicides into selective herbicides.

Bromoxynil is the only herbicide for which a herbicide-resistant transgenic plant has received European marketing consent (in 1996).

Interest in the transgenic approach for sulfonylurea has greatly decreased with the appearance of spontaneous resistance.

The Mechanisms of Resistance

Resistance to Amino Acid Biosynthesis Inhibitors

The enzymes for the synthesis of essential amino acids are a favoured target for the development of herbicides. These enzymes only exist in micro-organisms and plants. There has been a great deal of work carried out on herbicides of this type.

Glyphosate
(Target: 5-enolpyruvylshikimate 3-phosphate synthase EPSPS)

Since glyphosate was first described [4], the product *Roundup ready*, the isopropylamide salt of N-(phosphonomethyl)glycine, developed by Monsanto, has enjoyed a great deal of success. Although it is a total herbicide, with a dose of 1 kg/ha, it is the most widely used product in the world. It is especially used in intercultural situations or for perennial crops: 40% of the tonnage used in France is used during the rotation of annual crops (destruction of perennial weeds), another 40% to 45% is used on perennial non-rotated crops (vines, open spaces, etc.), the rest is used on fallow land (destruction or control).

Glyphosate can be used as a post-emergence weedkiller, several weeks before cultivation, by directed application on woody plants and even to weed uncultivated zones because the *in planta* mobility and stability of the product kills the most deeply rooted perennials.

The toxic effects on animals are relatively low. Although not very mobile in the soil and rapidly broken down by the microflora in the soil (half-life: 60 days), the product is chemically stable and not decomposed by the plants in which it circulates relatively quickly.

Selectivity

It is a total herbicide and, in spite of the widespread use, no cases of resistance have been reported to date. Only several biotypes of Indiana bindweed seem to manifest a certain tolerance although the mechanisms are still not known [5].

Target

Glyphosate reduces the synthesis of aromatic amino acids (phe, tyr, trp) (although their addition can eliminate the herbicide toxicity [6]) due to the competitive inhibition of 5-enolpyruvylshikimate 3-phosphate synthase (EPSPS) [7]. EPSPS, one of the best characterized enzymes in the biosynthesis of aromatic compounds, transfers the three phosphoenolpyruvate (PEP) carbons to shikimate 3-P.

Mode of Action

Bacterial and plant EPSPSs are monomeric enzymes. In plants, the enzyme is in chloroplasts although there also seems to be a cytosolic form [8]. Shikimate 3-P initially interacts with the enzyme, while PEP interacts preferably with the EPSPS-shikimate 3-P complex. This case is unique for enzymes where PEP is the substrate [9]. Glyphosate acts like a specific competitive PEP inhibitor and like an unspecific inhibitor of shikimate 3-P. Moreover, the herbicide does not interact with other enzymes that bind PEP [7].

In spite of the many studies devoted to EPSPS and to its interactions with glyphosate, there is still not enough information to determine the mode of action of the herbicide. In particular, the regulation of EPSPS in plants is still not fully understood (overexpression in the meristems and especially in the flowers, related to the role of the products of shikimate metabolism such as phytoalexins, lignins, ubiquinones, flavonoids).

Resistance Mutations and Molecular Data

Most of the resistance mutations, both in bacteria and in plants, provoke the overexpression of EPSPS [10]. In *Escherichia coli*, a substitution (gly96 to ala), in a very conserved part of the protein, reduces the glyphosate sensitivity by about 8,000, probably due to an affinity for PEP that is 13 times higher than that of the wild-type enzyme [11]. There is also a mutation in *Salmonella* (pro101 to ser) that gives glyphosate resistance, but it is not as effective [12].

Transgenes

The transfer of overexpressed genes [13] or mutant genes conferring increased glyphosate tolerance whose expression is directed by strong promoters [14] or mutant genes whose product is aimed at chloroplasts [15] has often turned out to be disappointing. Although the transgenic plants are always more tolerant than the controls, the level of protection is often inadequate for agricultural use. The growth of transgenic plants treated with glyphosate was in general slowed down, probably due to the accumulation of glyphosate in the meristems [10]. Most often, the normal growth of plants in the presence of glyphosate is only obtained by the transfer of two associated genes: one overexpressed gene, directed by a strong promoter, whose product is targeted to chloroplasts and a detoxification gene coding for bacterial glyphosate oxidoreductase [16] which breaks the glyphosate down into non-toxic aminoethyl-phosphonate and glyoxalate. However, good levels of resistance have recently been obtained using only one of the two genes described above.

As regards the management of the herbicide, the difficulty of selecting tolerant mutants, along with the lack of resistant adventitious plants is quite remarkable. This experimental finding is perhaps the result of very low frequencies of resistance mutations which, along with the low persistence in the soil, are very positive elements for its use. However, the impossibility of selecting resistant plants may also result from the fact that such mutations may be lethal. This is also a favourable element in the management of the use of the herbicide. If the low toxicity of glyphosate in animals is considered along with its low mobility due to its absorption and rapid degradation (breaking the carbon-phosphate bond) in surface soils, the portrait of the almost ideal herbicide is revealed.

Phosphinothricin (Target: Glutamine Synthetase {GS})

Two compounds have recently been developed and used as total herbicides: phoshinothricyl-alanyl-alanine, a tripeptide, for which the production by *Streptomyces hygroscopicus* has been mastered (marketed as bialaphos (*Herbiace*) by the Japanese company Meiji Seika Kaisha), and synthetic phosphinothricin (commercialized in the form of the ammonium salt of a mixture of D- and L-phosphinothricins (glufosinate-ammonium) under the trade name *Basta* by Hoechst. L-phosphinothricin is the active ingredient (the tripeptide is hydrolysed in living cells), and 1 to 4 kg/ha of product is applied to the leaves.

Phosphinothricin has been authorized since 1985 in about fifty countries. The life-span of the product in the soil and the toxicity are low.

Selectivity

Basta, a total post-emergence herbicide is recommended for established crops, especially ligneous crops, in nurseries or orchards (70% of its use) and to provoke drying in certain crops (potatoes). The recent and limited use of the product does not provide any information about the possibility of the appearance of resistant adventitious populations. No resistance has been reported to date.

Target

Phosphinothricin behaves like an analogue of glutamic acid and inhibits glutamine synthetase (GS). First described as an antibiotic [17], it was subsequently developed as a herbicide since it provokes the rapid accumulation of ammonium (toxic) [18].

Mode of Action

Phosphinothricin acts like a reversible GS inhibitor due to the greater affinity of the enzyme for this molecule than for glutamate (the first step in glutamine synthesis by GS forms the a-glutamylphosphate from glutamate and ATP) and it inactivates the enzyme in the presence of ATP [9]. The accumulation of ammonium is generally toxic for plants, especially in the chloroplasts. This is the main mode of action of the herbicide [18]. However, the exact mode of action is still not known since, for example, high ammonium levels (27 times as much) are not toxic for alfalfa cells grown *in vitro* [19].

Resistance Mutations and Molecular Data

Plant GSs are generally proteins with 8 subunits. There are different plant GS isozymes, coded by different nuclear genes. They are distinguished according to their structure, location, role and regulation [21]. GS can be divided into two main groups: cytosolic GS1 coded by three or four genes depending on the plant species, and chloroplastic GS2. This distribution may be due to the essential role of glutamine in the synthesis of other amino acids or by the toxicity of the ammonium ion that requires rapid assimilation, whatever the metabolic conditions in each organ. This complex and multigenic system of regulation also accounts for the fact that no GS1 mutants have been isolated except for herbicide-resistant alfalfa cells [21] resulting from the amplification of a GS1 gene.

Transgenes

As in the case of glyphosate resistance, the transfer of amplified alfalfa genes to tobacco did not confer usable levels of resistance [22]. Both the number of GS genes and the apparent complexity of their regulation make it extremely difficult, if not impossible, to significantly modify their expression. The search for an effective resistance thus turned towards detoxification. The *bar* gene from *Streptomyces* [23] has provided an effective system of phosphinothricin detoxification for plants after modification of the initiation codon [24]. Similar enzymes (phosphinothricin acetyl-transferase, PAT) derived from *S. viridochromogenes* have been used for the same purpose [25].

Chimaeric constructs including the strong "35S" promoter from cauliflower mosaic virus (CaMV) directing the expression of bacterial *bar* or *pat* genes (where the codons have been modified for optimum expression in plants) have been used to transfer *Basta* resistance to very different plants: tobacco, tomato, lettuce, maize, rape, etc. In certain cases, the agronomic validation of the resistance has been found to be fully satisfactory.

Conclusion

The development of this strategy is a good example of the possibilities of genetic engineering. Providing that the rapidity of degradation and the ecological safety of the compound (as well as the acetylated form derived from its detoxification) are verified, glufosinate may turn out to be an excellent herbicide. Management of the herbicide will be facilitated if the absence of natural resistance mutations is verified under field conditions. As with glyphosate, it is necessary to avoid the risk of releasing resistance genes.

Resistance to Photosynthesis Inhibitors

Photosynthesis, limited to certain bacteria, algae and green plants is also a choice target for herbicides. Excluding products that globally interfere with photosynthesis, affecting either membranes or pigment synthesis, it is remarkable to note that most herbicides in this category are specific for one of two precise targets. The first group affects the transfer of electrons from plastoquinone Qa to plastoquinone Qb by modifying protein D1, bound to Qb and located in the chloroplast membrane (photosystem II). The second group traps electrons from photosystem I destined for ferredoxin, by releasing toxic superoxide radicals.

Bromoxynil (Target: Photosystem II)

Bromoxynil is a p-hydroxybenzonitrile compound. It has been commercialized in Europe as a herbicide since 1963 by May and Baker (trade name: *Buctril*). Bromoxynil is rarely used alone (as for example 600 g/ha on maize) but rather on other cereals combined with oxynil or auxins. This product is sold in about 46 countries and is registered for use on monocots.

Selectivity

Cereals convert bromoxynil into compounds that are one hundred times less toxic for the plant in three successive stages: the hydrolysis of the nitrile group, followed by the decarboxylation into dibromophenol and the replacement of the bromine atoms by hydroxyl groups, thereby releasing the phenols [26]. Most dicots are sensitive and thereby eliminated.

Target

In sensitive plants, the mortality results from the inhibition of the transport of photosystem II electrons.

Mode of Action

Like other herbicides of this type, bromoxynil competes with plastoquinone Qb (an electron carrier) for binding to the D1 protein (32 kDa) of the chloroplast membrane [27]. The sites for the fixation of bromoxynil differ from those of triazines [28].

Resistance Mutations and Molecular Data

The selected *in vitro* resistance in bacteria results from mutations at the benzonitrile fixation sites on the D1 protein [28]. The gene sequence (psbA) from this membrane protein, coded by the chloroplast, turns out to be very much conserved from cyanobacteria to higher plants [29].

Transgenes

Nitrilase genes detected in soil bacteria (*Klebsiella ozaenae*, for example) [30] were chosen for transfer to dicots [31].

The chimaeric genes used carry the nitrilase coding sequence of *Klebsiella*, modified for use in the plant. The expression is either directed by a 35S promoter from CaMV or by the promoter of the gene of the small subunit of Rubisco from sunflower. Transferred to tobacco, these constructs allow the plants to develop normally, even with very high doses (about 5 kg/ha) of active ingredient [31]. The agricultural efficacy of these same constructs is being assessed, mainly in rape.

Conclusion

As with the triazines (see below), more intensive use of the product may favour the appearance of resistant adventitious plants by mutation of the product target. Resistant *Chenopodium* have already been described [32].

Different Resistances Being Currently Assessed

The strategies for the creation (or transfer) of resistances that are currently in the assessment process today involve all families of herbicides, as indicated by the following examples.

Auxin: 2,4-D and Derivatives (Target Not Specified)

Synthetic auxin analogues have been used for the last forty years to control post-emergence dicots in monocot crops (maize, wheat). Their targets are not yet known. The difference in sensitivity between the two types of plants probably results from the metabolic differences concerning auxins as well as cytochrome P-450 monooxygenase activities. Surprisingly, considering the widespread use of these herbicides, and although the toxic action varies, no real problems of resistance have been noted to date.

The strategy chosen by several teams relies on the transfer of a 2,4-D mono-oxygenase from *Alcaligenes eutrophus*. This detoxification approach was first tested on tobacco [33,34] and has just been applied to cotton [35].

Triazines (Target: Photosystem II)

This category includes a series of compounds typified by atrazine. Like bromoxynil, they are competitive inhibitors of the association of the D1 protein (*psbA* gene) with the electron carrier Qb [27]. These compounds are used on maize crops. Maize is resistant due to a very active gluthatione-S-transferase that gives rise to inactive triazine conjugates [36]. Intensive and repeated use for the last forty years, combined with the persistence of the products, has given rise to a large number of resistant adventitious populations (57 different species on over one thousand sites) [37]. Most of the resistances are derived from mutations in the chloroplast D1 protein. More than twelve mutants from different organisms (*Chlamydomonas, Euglena, Anacystis* and plants) have been characterized. Over two-thirds are obtained from the substitution of the residue ser264 by gly or ala. This is the only mutation described in plants. However, another resistance, of non-maternal heredity, was noted in *Abutilon theophrasti* due to an effective conjugation of triazines by glutathione-S-transferase [38].

The strategies for the transfer of triazine resistance differ greatly. One consists of the transfer, either by cross-fertilization, or by protoplast fusion, of the resistant chloroplast from *Brassica campestris* to a cultivated turnip or rape [39]. The same transfer approach can be applied between all resistant adventitious plants and cultivated plants of the same family (*Solanaceae* and potatoes, for example) [40]. It is also possible to select *in vitro* resistance mutations, provided that it is possible to establish photosynthetic growing conditions that allow for the use of the appropriate selective screen [41].

Another strategy, employing the transgenic approach, consists in cloning the *psbA* gene from a resistant adventitious plant and then transferring it to the nuclear genome of the plant of agronomic interest endowed with a sequence encoding a chloroplast targeting signal [42]. Finally, it is also possible to transfer the gluthatione-S-transferase gene that is effective in the inactivation of triazines by conjugation [22].

Bipyridil (Paraquat) (Target: Photosystem I)

Paraquat and diquat total herbicides have been used extensively for the last thirty years. Their toxicity is due to the formation of very reactive free radicals [43]. Very dangerous for animals, they are not very persistent since they migrate little in the soil and are quickly destroyed by micro-organisms. Paraquat is not very active on young cereals due to the optimum regulation of the three activities that are required for defence against free radicals: glutathione reductase, ascorbate peroxidase and superoxide dismutase.

One of the strategies being studied for the creation of resistance consists in the transfer of superoxide dismutase genes to cultivated plants [44].

Weeding in Major Crops

Maize: Practically the entire planted surface is treated with herbicide before germination of the crop (atrazine still accounts for 80% of these basic treatments in 1996). The post-germination treatments are considered to be corrective treatments mainly in the fight against triazine-resistant species.

Rape: Practically all of the surfaces planted with rape are weeded before germination with formulations consisting of trifluraline, metazachlore and a mixture of tebutame + clomazone. Post-germination herbicides are only used on less than half of the rape surfaces to specifically fight grassy weeds or certain dicots, rarely against all flora.

Beet: The pre-germination treatment with a mixture of metamitrone + chloridazone, applied to 80% of planted surfaces, is systematically followed by 2 or 3 post-germination treatments. These post-germination treatments, carried out with a mixture of phenmediaphane + metamitrone on all beet crops, are completed with specific anti-grass treatments.

Chicory: For chicory, pre-germination weeding is systematic (benfluarine, propyzamine and chloprophame). There are few post-germination herbicides and they are not used very much.

Tomato: The tomato also undergoes pre-germination treatment with both seeded or replanted crops. This is followed by a sometimes repeated post-germination treatment with metribuzine.

Potato: Pre-germination treatment with metribuzine is followed by post-germination treatments at the beginning of vegetation (diquat) or later with metribuzine and anti-grass treatments.

Tobacco: There are two specific cases. Tobacco is weeded mechanically or chemically (pre-germination using metobromuron and pendimethaline; post-germination using anti-grass treatments). It benefits from mechanical maintenance limiting the growth of perennials.

Melon: Given its sensitivity to herbicides, melon can only be treated with chlorthal, an active ingredient with a limited action spectrum.

The weed problem is fairly well resolved with the chemical means available for most of the crops mentioned (except for melon and, to a lesser extent, tobacco). However, the cost of weeding is a major problem today. By way of example, the average cost for rape is 450 francs/hectare. The cost exceeds 1,000 francs in certain situations.

Transgenic plants may change the technical choices for weeding by developing post-germination treatments and by replacing persistent and/or toxic herbicides by non-persistent and/or less toxic herbicides. There may also be a reduction in the cost of treatment.

Consequences of the Large-Scale Use of Herbicide-Resistant Transgenic Plants

The main problem raised by herbicide-resistant plants is to make sure that the resistance character does not spread and become an environmental problem and especially a problem for the farmer. Everything depends on the plant species (tendency to regrow the following year, seed dormancy, range of dissemination of pollen), the type of herbicide and especially the farming practices.

Gene Dissemination

Most European plants (cereals, tobacco, etc.) are geographically isolated from their wild relatives and are not interfertile with them. Therefore the problem of gene flow does not exist. However, this is not the case with certain species such as rape, beet or chicory (refer to the chapter on gene flow). For these three species, it is highly likely that the transgenic resistance will finally become established in a wild related species having a certain interfertility with the transgenic variety [45]. Current opinion varies as to the extent and rapidity of the phenomenon.

The Answers Provided by Population Genetics

The results of the local models developed by P.H. Gouyon's team (Orsay School of Sciences) do not currently allow exact predictions of the spread of a (dominant) transgene in the population either on a short-term (after five generations) or a long-term basis (equilibrium).

Short-term predictions are difficult due to the great many interactions between the different factors. Since the effect of one factor on the frequency of the transgene in the population is determined by the value of the other factors, it is necessary to know the exact value of all of the interacting factors in order to determine the frequency of the transgene. This is not always possible.

On a long-term basis, the determinist models indicate that only one factor is generally sufficient to predict the frequency of the transgene. However, rare events, in time or space, but resulting in a strong migration, may totally modify the development of the transgene. Reliable predictions are not possible due to the unpredictability of these effects.

However, the models allow different scenarios to be envisaged ranging from a very low dissemination of the transgene to the rapid invasion of the wild population. In order to determine the most probable outcome for a plant/transgene pair, it is necessary to know the value of certain factors. This is not available for most pairs that are candidates for commercialization.

According to the work carried out by the same team, the regional models show that in the theoretical cases studied, it is possible to expect a certain speed of propagation of resistant adventitious species. However, the real situation is complex and the land is more varied than that considered by these models.

Assessment of the risk is still an open question and most likely will only be resolved by collecting data, at the field level, during the large-scale cultivation of herbicide-resistant transgenic plants.

The models currently available are not able to provide a quantitative answer. However, herbicide resistance characters from transgenic plants will probably spread through inter-fertile populations, due to their negligible "cost" to the plants.

In how many years? Over what area? These questions still have to be answered.

Consequences on Herbicides and their Uses

One of the obvious consequences is that herbicides that were "total" are no longer so, although they could have remained so for a long time.

Where the herbicides are not used, the character is neutral and does not present an additional problem.

However, the situation will change in zones where the herbicide is and will be used. The "total" herbicide will no longer be able to weed the treated area. For example, in an orchard where glyphosate is currently being used, a resistant rape may multiply without any competition.

Each situation will be specific. A resistant annual beet is a "weed" for cultivated beets and will no longer be eliminated by the herbicide that was chosen to destroy it in the first case.

Solutions proposed by experts are indicated in *Table I* for each pair (*i.e.* beet/total herbicide and rape/total herbicide) identified as potentially problematic in the rotation of crops.

Table I. Destruction of volunteers: active ingredients or techniques (according to the Organisation Nationale Interprofessionnnelle des Bio-industries).

Tolerant volunteers / **Crops**	**Beet**	**Rape**	**Maize**
Cereals	Many anti-dicot herbicides	Many anti-dicot herbicides	NOT CONCERNED
Peas	Aclonifen Fluorochloridone Bentazone	Aclonifen Fluorochloridone Bentazone	
Sunflower	Aclonifen Fluorochloridone	Aclonifen Fluorochloridone	
Maize	Rare (Triazines)	Triazines Pyridate Sulfonylureas Sulcotrione	
Potato	Aclonifen Fluorochloridone Substituted ureas	Aclonifen Fluorochloridone	
Flax	MCPA	MCPA	
Beet	No change - mechanical weeding - manual weeding	Metramitrone BTGV Trisulfuron	
Rape	(Colzor) (Butisan) (Treflan) Rare	None (as currently) - destruction after harvest - false-seedings	
Fallow	Sulfonylureas Hormones Mechanical crushing	Sulfonylureas Hormones Mechanical crushing	

It should be noted that the recommended solutions are currently available. According to the professionals, another total herbicide is not likely to be marketed in a short or medium term.

Another important factor is that the extension of the herbicide/herbicide-resistant transgenic plant pairs are based on "friendly" herbicides (low life-span, low toxicity, high efficiency). They have replaced older and less ecologically "friendly" herbicides. There are therefore short-term ecological benefits. It is necessary to see whether this isn't counterbalanced on a long-term basis by the fact that these "friendly" herbicides will lose their total herbicide character. It is very difficult to decide without a global ecological assessment taking all the factors into account.

In addition, as described above, some of these herbicides will be broken down by the genes introduced into the plant. The new metabolites obtained have to be precisely assessed as regards their potential effects on public health and the environment. The Biomolecular Engineering Commission and the Toxic Substances Commission are in the process of assessing bromoxynil, glyphosate and phosphinothricin.

Conclusion

Herbicide-resistant transgenic plants have become of special importance in the debate, not because they involve a serious risk, but because of the great number of cultivated varieties and the huge surfaces involved. The massive aspect of their potential use, or even the possible banalization of these genetically modified organisms, give special weight to the decisions that have to be taken regarding their commercialization.

It is illusory to imagine that the resistant transgenic plant/herbicide pair will be enough to control weeds. New combinations of treatments will certainly be required.

The most obvious short-term economic benefits are expected with chicory and beets and to a lesser extent with rape.

The main question in this context lies more in the lack of knowledge of the range of products that will be used in fifteen or twenty years. In fact, most herbicide authorizations covering the oldest and potentially most dangerous products that are not covered by patent are being reexamined, especially in Europe. Nothing today indicates that new products will be put to use. Considering the time required nowadays for product development, new formulations will probably not be numerous. The range of products available will not increase; on the contrary, it may decrease.

Within this context, certain major total herbicides may disappear and become selective treatments.

It is difficult to predict the cost/benefit long-term ratio quantitatively. If the immediate ecological benefit is clear, the long-term "loss" in efficacy of the herbicides seems inevitable, and assessment is difficult.

In fact, no study is reliable enough to predict the time and surfaces involved. A network will have to be set up to monitor the consequences of the large-scale dissemination of transgenic plants and describe the progression of resistance from year to year.

This will constitute a warning system in case of even a local explosion of the phenomenon. The factors to monitor and the operation of this system will quickly have to be determined by population geneticists, agronomists, professionals and the public authorities.

References

1. Christianson ML. Fun with mutants: applying genandic mandhods to problems of weed physiology. *Weed Sci* 1991 ; 39 : 489-96.
2. Anderson PC, Georgeson M. Selection and characterization of imidazolinone tolerant mutants of maize. In : *Biochemical Basis for Herbicide Action.* Ashford, UK 1986: 27th Harden Conf Prog. Wye College (Abstract).
3. Arnould, *et al.* OGM : une théorie pour les risques. *Biofutur* 1993 ; 6 : 45-50.
4. Baird D, Upchurch R, Momesley W, Frantz J. Introduction of a new broad spectrum postemergence herbicide class with utility for herbaceous perennial weed control. *Proc North Cent Weed Control Conf* 1971 ; 26 : 64-8.
5. Degennaro FP, Weller SC. Differential susceptibility of field bindweed (*Convolvulus arvensis*) biotypes to glyphosate. *Weed Sci* 1984 ; 32 : 472-6.
6. Jaworski EG. Mode of action of N-phosphomethylglycine inhibition of aromatic aminoacid biosynthesis. *J Agric Food Chem* 1972 ; 20 : 1195-8.
7. Steinrucken H, Amrhein N. The herbicide glyphosate is a potent inhibitor of 5-enolpyruvyl-shikimate acid 3-phosphate synthase. *Biochem Biophys Res Comm* 1980 ; 94 : 1207-12.
8. Moudale D, Coggins J. Subcellular localization of the common shikimate pathway enzymes in *Pisum sativum L. Planta* 1985 ; 163 : 241-9.
9. Kishore GM, Shah DM. Aminoacid biosynthesis inhibitors as herbicides. *Annu Rev Biochem* 1988 ; 57 : 627-63.
10. Kishore GM, Padgette SR, Fraley RT. History of herbicide tolerant crops, methods of development and current state of the art - emphasis on glyphosate tolerance. *Weed Technol* 1992 ; 6 : 626-34.
11. Kishore GM, Brundage L, Kolk K, Padgette S, Rochester D, Huynh Q, Della-Cioppa G. Isolation, purification and characteristic of a glyphosate tolerant mutant of *E. coli* EPSP synthase. *Fed Proc Am Soc Esp Biol* 1986 ; 45 : 1506.
12. Comai L, Sen L, Stalker D. An altered *aroA* gene product confers resistance to the herbicide glyphosate. *Science* 1983 ; 221 : 370-1.
13. Shah DM, *et al.* Engineering herbicide tolerance in transgenic plants. *Science* 1986 ; 233 : 478-81.

14. Comai L, Facciotti D, Hiatt W, Thompson G, Rose R, Stalker D. Expression in plant of a mutant *aroA* gene from *Salmonella thyphimurium* confers tolerance to glyphosate. *Nature* 1985 ; 317 : 741-4.
15. Della-Cioppa G, Bauer S, Taylor M, Rochester D, Klein B, Shah D, Fraley R, Kishore G. Targeting a herbicide resistant enzyme from *Escherichia coli* to chloroplasts of higher plants. *Bio/Technology* 1987 ; 5 : 579-84.
16. Pipke R, Amrhein N. Isolation and characterization of a mutant of *Arthrobacter* sp. strain GLP1 which utilizes the herbicide glyphosate as its sole source of phosphorus and nitrogen. *Appl Environ Microbiol* 1988 ; 54 : 2868-70.
17. Bayer E, Gugel KH, Hagele K, Hagenmaier H, Jessipow S, Konig WA, Zahner H. Stoffwechselprodukte von Microorganismen phosphinothricin und phosphinothriciyl-alanyl-alanin. *Helvetica Chim Acta* 1972 ; 55 : 224-39.
18. Tachibana K, Watanabe T, Sekizawa Y, Takenmatsu T. Accumulation of ammonia in plants treated with bialaphos. *J Pestic Sci* 1986 ; 11 : 33-7.
19. Krieg LC, Walker MA, Senaratna T, McKersie BD. Growth ammonia accumulation and glutamine synthetase activity in alfalfa shoots and cell cultures treated with phosphinothricin. *Plant Cell Rep* 1990 ; 9 : 80-3.
20. McNally SF, Hirel B, Gadal P, Mann F, Stewart GR. Glutamine synthetases of higher plants. *Plant Physiol* 1983 ; 72 : 23-5.
21. Donn G, Tisher E, Smith JA, Goodman HM. Herbicide resistant alfalfa cells: an example of gene amplification in plants. *J Mol Appl Genet* 1984 ; 2 : 621-35.
22. Mazur BJ, Falco SC. The development of herbicide resistant crops. *Ann Rev Plant Physiol Plant Mol Biol* 1989 ; 40 : 441-70.
23. Thompson CJ, Movva NR, Tizard R, Crameri R, Davies JE, Laurwereys M, Botterman L. Characterization of the herbicide resistance gene bar from *Streptomyces hygroscopicus*. *EMBO J* 1987 ; 6 : 2519-23.
24. Deblock M, Botterman J, Vandewiele M, Docks J, Thoen C, Gossele V, Movva NR, Thompson C, Van Montagu M, Leemans J. Engineering herbicide resistance in plants by expression of a detoxifying enzyme. *EMBO J* 1987 ; 6 : 2513-8.
25. Wohlleben W, Arnold W, Broer I, Hillemann D, Strauch E, Puhler A. Nucleotide sequence of the phosphinothricin N-acetyl transferase gene from *Streptomyces viridochromogenes* Tu 494 and its expression in *Nicotiana tabacum*. *Gene* 1988 ; 70 : 25-37.
26. Buckland JL, Collins RF, Pullin EM. Metabolism of bromoxynil octonoate in growing wheat. *Pestic Sci* 1973 ; 4 : 149-62.
27. Vermaas W. Molecular-biological approaches to analyse photosystem II structure and function. *Ann Rev Plant Physiol Plant Mol Biol* 1993 ; 44 : 457-81.
28. Ajlani G, Meyer I, Vernotte C, Astier C. Mutation in phenol-type herbicide resistance maps within the gene *psbA* in *Synechocystis* 6714. *FEBS Lett* 1989 ; 246 : 207-10.
29. Zurawski G, Bohnert H, Whitfield P, Bottomley W. Nucleotide sequence for the M 32000 thylakoïd membrane protein from *Spinacia oleracea* and *Nicotiana debneyi* predict a totally conserved primary translation protein of M 38950. *Proc Natl Acad Sci USA* 1982 ; 79 : 7699-703.
30. Stalker DM, McBride KE. Cloning and expression in *Escherichia coli* of a *Klebsiella ozaenae* plasmid borne gene encoding a nitrilase specific for the herbicide bromoxynil. *J Bacteriol* 1987 ; 169 : 955-60.
31. Leroux B, Lebrun M, Garnier P, Sailland A, Pelissier B, Freyssinet G. Engineering herbicide resistance in tobacco plants by expression of a bromoxynil specific nitralase. *Bull Soc Bot Fr* 1990 ; 137 : 65-78.
32. Holt JS, Lebaron HM. Significance and distribution of herbicide resistance. *Weed Technol* 1992 ; 4 : 141-9.
33. Lyon BR, *et al.* Expression of a bacterial gene in transgenic tobacco plants confers resistance to the herbicide 2,4-dichlorophenoxyacetic acid. *Plant Mol Biol* 1989 ; 13 : 533-40.

34. Streber WR, Willmitzer L. Transgenic tobacco plants expressing a bacterial detoxifying enzyme are resistant to 2,4-D. *Bio/Technology* 1989 ; 7 : 811-6.
35. Lyon BR, Cousins YL, Llewellyn DJ, Dennis ES. Cotton plants transformed with bacterial degradation gene are protected from accidental spray drift damage by the herbicide 2,4-D. *Transgenic Res* 1993 ; 2 : 162-9.
36. Edwards R, Owen WJ. Comparison of gluthation-S-transferase of *Zea mays* responsible for herbicide detoxification in plants and suspension cultured cells. *Planta* 1986 ; 169 : 208-15.
37. Holt JS, Powles SB, Holtum JA. Mechanisms and agronomic aspects of herbicide resistance. *Ann Rev Plant Physiol Plant Mol Biol* 1993 ; 44 : 203-29.
38. Anderson MP, Gronwald JW. Atrazine resistance in a velveleaf (*Abutilon theophrasti*) biotype due to enhanced gluthatione S-transferase activity. *Plant Physiol* 1991 ; 96 : 104-9.
39. Beversdorf WD, Weiss-Lerman J, Erickson LR, Souza-Machado V. Transfer of cytoplasmically-inherited triazine resitance from bird's rape to cultivated *Brassica campestris* and *B. napus. Can J Genet Cytol* 1980 ; 22 : 167-72.
40. Austin S, Helgeson JP. Interspecific fusion between *Solanum brevidens* and *S. tuberosum. Plant Mol Biol* 1987 ; 140 : 209-22.
41. Fluhr R, Cseplo A. Induction and selection of chloroplast coded mutations in *Nicotiana. Methods Enzymol* 1986 ; 118 : 611-23.
42. Cheung AY, Bogorad L, Van Montagu M, Schell J. Relocating a gene for herbicide tolerance: a chloroplast gene is converted into a nuclear gene. *Proc Natl Acad Sci USA* 1988 ; 85 : 391-5.
43. Dodge AD. Herbicide interacting with photosystem I. In : Dodge AD, ed. *Herbicides and Plants Metabolisms*. Cambridge Press, 1989 : 37-50.
44. Bowler C, van Montagu M, Inze DD. Superoxyde dismutase and stress tolerance. *Ann Rev Plant Physiol Plant Mol Biol* 1992 ; 43 : 83-116.
45. Darmancy H. The impact of hybrids between genetically modified crop plants and their related species: introgression and weediness. *Mol Ecol* 1994 ; 3 : 37-40.

9

Virus-Resistant Transgenic Plants

Hubert Laude*

Summary

During the last ten years, transformation has proven to be a particularly powerful means of providing cultivated plants with resistance towards one or several viral pathogens. The most promising approach to date consists of introducing a transgene derived from the target virus itself into the nuclear genome of the plant. The expression of a transgene encoding a viral capsid protein is a strategy which has been successfully applied to a large number of host/virus systems. Transgenic expression of other viral genes (sometimes modified), as well as the expression of non-coding, sense or antisense viral transcripts, has also provided virus-resistant plants. The aim of this chapter is to evaluate the possible environmental impact of introducing such plants into agricultural practice, in terms of the evolution of viral populations or the spread of transgene-associated resistance. Effects already observed in non-transgenic plants, such as trans-capsidation, recombination or mutation may also occur in infected transgenic plants expressing viral sequences, thus possibly creating a transient or stable modification in a given viral population. However, the chances of this happening appears to be low. Technical improvements that would reduce the probability of dissemination of the expressed viral

* Member of the CGB, Senior Scientist, INRA, Jouy-en-Josas, France.

material can be envisaged. A case-by-case assessment of every new host/transgene pair, combined with monitoring, whenever necessary, of the virus-resistant variety during the early phase of development, should prevent any unexpected harmful effects.

Viral diseases that strike valuable crop plants result in large economic losses. The development of genetically resistant plants is one way to control these infections. The standard methods of selection, in particular crossing with wild related varieties, have introduced resistance into different cultivated varieties. The recognition of a pathogen by a plant is determined by certain genes from the host, called PR or pathogenesis-related genes. They coordinate the development of an active response in the infected host. The modulation of the expression of such genes *via* transformation may, in the future, lead to the creation of virus-resistant plants. However, the identification of the genes involved in natural virus resistance is just beginning [1].

Current approaches differ from natural resistances because they introduce genetic information obtained from the target virus into the genome of the plant. The resulting resistance to infection is of relatively restricted specificity although it may be very strong. The general mechanisms underlying this resistance are only partly known. A direct harmful effect of the uncoordinated expression of viral information on the reproduction or propagation of the attacking virus, *i.e.* a passive defence, is the mechanism most often predicted although it is probably not the only one. Whatever the case, the research carried out during the last decade has largely confirmed the possibility of creating plants with specific resistance or tolerance by integrating viral sequences.

The expected benefits of cultivating these transgenic plants is such that they will most likely be introduced into agriculture in the near future. Out of about one thousand transgenic plant trials in unconfined environments carried out in the world between 1987 and 1996, about 12% involve viral genes. This chapter describes the strategies used and the potential environmental risks that have to be taken into account in the assessment of the cost/benefit ratio for such plants.

The Techniques Involved

The strategies aiming at making a plant resistant towards a specific virus or a group of related viruses involve a transgene expressing either a standard coding or non-coding transcript or a transcript to be co-replicated by the infecting virus. The approaches being envisaged are continually evolving. This is more an indication of the multiplicity of infecting viruses than the number of target plants.

Resistance by Expression of a Viral Protein

In most plant viruses, the virions consist of RNA surrounded by a capsid formed by the assembly of a single protein [2]. The approach based on the expression of the capsid gene (coat protein-mediated resistance) is currently considered as the most promising since it has already been successfully applied to over forty RNA viruses belonging to at least ten distinct taxonomic groups infecting a fairly wide variety of dicots and monocots [3,4].

The conceptual simplicity of the approach, illustrated in *Figure 1*, should not mask the diversity, or even the plurality of the mechanisms involved. However, it is not necessary to go into this here, apart from noting that this is likely to be due to the multiplicity of functions that the capsid protein has during the viral cycle, in replication, the movement of virions, the recognition by host resistance factors and transmission by the vector [5].

A potential undesirable effect related to the use of such a strategy is heterologous encapsidation or transcapsidation, *i.e.* the possibility that part or all of a non-homologous viral genome becomes encapsidated by the capsid protein synthesized from the transgene. Encapsidation relies on two types of stereo-specific interactions: the recognition by the capsid protein of an encapsidation signal consisting of one or several sequences of viral nucleic acid, and protein/protein interactions. Therefore, heteroencapsidation is in principle disfavoured during the co-infection of a plant by two unrelated viruses.

Heteroencapsidation is plausible or even expected when the viral entities are genetically close, or strains from the same virus. Experiments have shown that complementation of one strain of transmission-defective potyvirus can occur following the infection of a transgenic plant expressing a capsid protein derived from a strain that is transmissible by aphids [6]. Heteroencapsidation may thus allow a viral strain to spread by a novel route. Such a modification only involves one generation since the genetic material from the heteroencapsidated virus does not change in this case. A specific case in which the risk related to hetero-encapsidation seems to be tangible [7] is that of agents such as umbraviruses. These rely on a virus from another group for encapsidatation and transmission.

A technical solution to mitigate the problems related to heteroencapsidation is based on the alteration, in the capsid protein, of the region involved in the transmission by the vector. Different greenhouse or field tests have demonstrated that the expression of such a disarmed protein may not have any significant effect on the resistance induced [8]. Moreover, the expression of genes coding for proteins other than the nucleocapsid protein, especially non-structural proteins, is an alternative that is promising even if it cannot be generalized. Solid resistance to infection has been reported in plants expressing a dysfunctional form of the viral replicase [9] (*Figure 1b*) or a protein necessary for the cell-to-cell movement of virions [10].

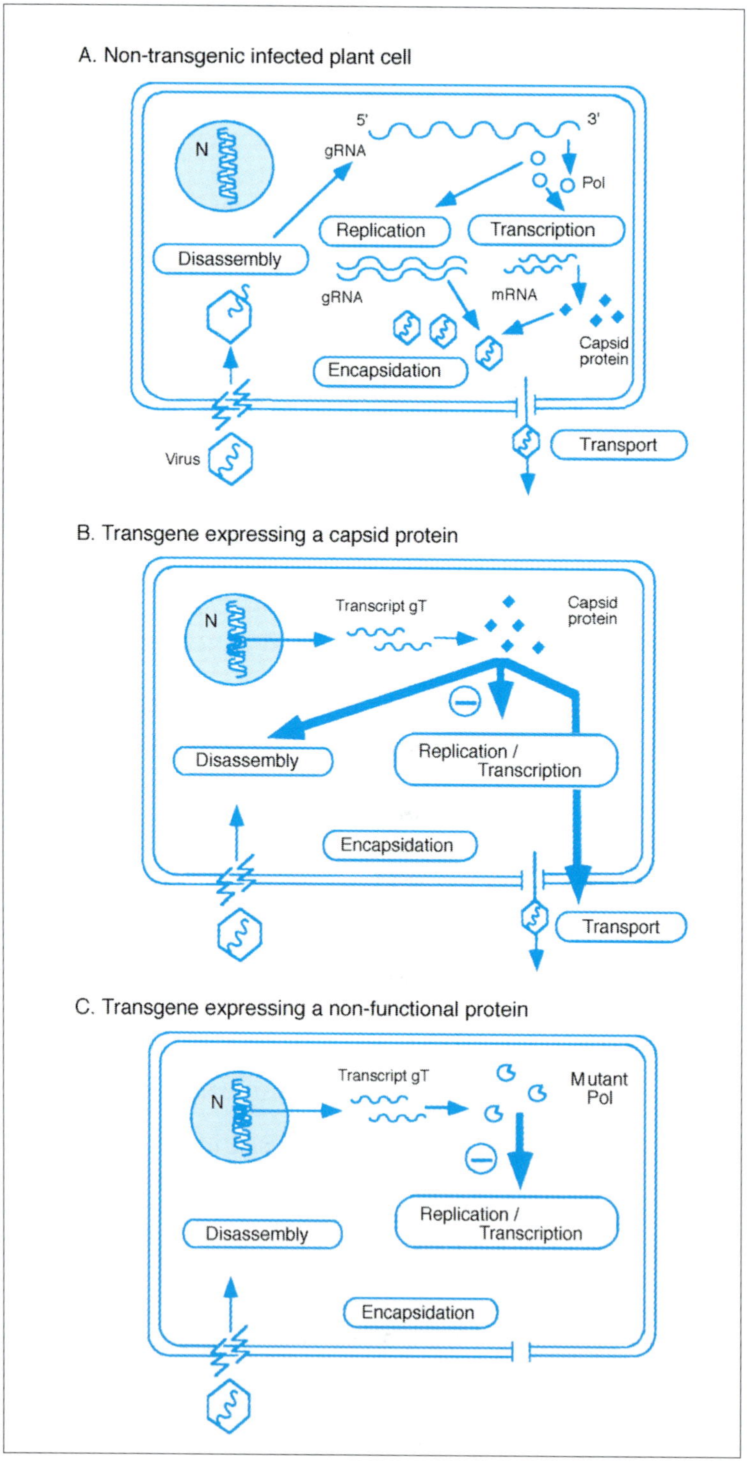
A. Non-transgenic infected plant cell
5'
3'
N
gRNA
Pol
Replication
Transcription
Disassembly
gRNA
mRNA
Capsid
protein
Encapsidation
Transport
Virus
B. Transgene expressing a capsid protein
N
Transcript gT
Capsid
protein
Disassembly
Replication /
Transcription
Encapsidation
Transport
C. Transgene expressing a non-functional protein
N
Transcript gT
Mutant
Pol
Replication /
Transcription
Disassembly
Encapsidation

Figure 1. Resistance by expression of a viral protein.

An example of a viral cycle is illustrated in **A**. The infecting virus penetrates the cell (vector, wound). The virion is disassembled and releases its genomic RNA (gRNA). In the chosen example (representative of most plant viruses), the viral RNA polymerase (Pol) is directly translated from the gRNA. This enzyme transcribes messenger RNA (mRNA) including that of the capsid protein, and synthesizes new gRNA. The latter is encapsidated and the virions formed infect other cells, by cell-to-cell transfer and long-distance transport, and possibly other plants. **B.** The transgene integrated in the plant genome codes for a capsid protein which is constitutively synthetized and accumulates in the cell. This protein may specifically inhibit different stages in the viral cycle, depending on the virus family. In the case of the TMV virus (tobamovirus), one of the best documented, the inhibition is very early and mainly involves the virion disassembly stage (the resistance may be overcome by the direct introduction of gRNA into the cell). In other viruses, the inhibition comes much later and involves the synthesis of viral RNA (*e.g.* replication/transcription imbalance), or even the movement of viral particles when the capsid protein is involved in this process. **C.** In this example, the transgene codes for a mutant form of viral polymerase. The dysfunctional polymerase exerts a dominant negative effect on the viral polymerase coded by a homologous infecting virus. Very solid resistance based on this mechanism has been described in the case of PVX (potexvirus). In cases B and C, the viral progeny arising from the first infected cells will be reduced or absent. This avoids the spread of the infection to surrounding tissue. The resistance observed is mainly passive, that is, it does not rely on the activation of local or systemic defences of the plant by the product of the transgene.

Another potential risk associated with the expression of a coding or non-coding viral transcript, that is involving recombination with the genome of a related virus, will be considered in the following section.

Resistance by Expression of a Non-Coding Transcript

Although not as well documented as the previous technique, this approach consists in expressing an inactive sense or antisense transcript and appears to work for a certain number of plant/virus pairs [3]. The possibility of controlling the infection by means of a viral transcript (RNA-mediated resistance) was demonstrated by the analysis of a series of control plants, transformed by a non-coding capsid protein gene. The induced resistance may equal that observed with a translated transcript, demonstrating that the product coded by the transgene is not essential. Viral genome sequences other than those of the capsid gene may confer resistance. However, such an approach cannot be generalized as easily as one using a coding transcript.

The mechanism of resistance *via* the expression of viral sense RNA remains fairly conjectural. Interference of replication or translation resulting from RNA/RNA pairing has been suggested. In potyviruses, a separate mechanism has recently been proposed, related to the cosuppression phenomenon already observed in other types of transgenic plants. This model proposes that viral infection induces endogenous ribonucleases that selectively break down homologous viral sequences [11] (*Figure 2a*). The antisense RNA approach (*Figure 2b*), generally considered less effective, may be improved by the addition of ribozyme sequences to the transgene. The anticipated effect is a catalytic degradation of the target viral RNA [4].

The risk potentially associated with such strategies involves recombination between the transcript derived from the transgene and the viral genomic RNA. This recombination phenomenon has been well established in plant and animal viruses. The mechanism put forward is based on the propensity of viral RNA polymerase to pause during elongation, which may allow a switch in template. Compared to the fairly frequent production of defective RNA, derived from a homologous but aberrant recombination, legitimate homologous intergenomic recombination is only observed at an easily detectable frequency in certain families of viruses [12].

The emergence of a recombinant virus assumes that it can spread in the viral population, implying a distinct selective advantage. However, the emergence of viable recombinants following homologous transgene/virus recombination has until now only been demonstrated in conditions of strong selective pressure [13-15]. Although it is true that homologous RNA/RNA recombination is one of the factors that contributes to the natural genetic evolution of viral populations, the appearance of new virulent phenotypes in the field is most often due to mutations.

Heterologous recombination, and *a fortiori* gene capture (implying illegitimate double recombination), are expected very infrequently, since in this case RNA and not DNA is involved [12]. There do not seem to be any published data on the detection of recombination between viral genomic RNA and non homologous cellular RNA under experimental conditions (excluding the specific case of retroviruses). It is true that the genomic organization of several viruses strongly pleads in favour of the incorporation of an RNA sequence originating from the host cell during their evolution. In two animal viruses, this event was accompanied by a significant alteration in virulence or host range [16,17]. The only known example in plants involves a satellite RNA [18]. In the cases mentioned, it should be noted that the phenotypic change appears to depend more on the site of integration in the viral genome than on the type of sequence integrated. As a general rule, the addition of RNA derived from a transgene to the cellular mRNA pool should not present a truly novel situation in terms of illegitimate recombination.

Recombination phenomena are by nature not very controllable. However, different technical improvements may be considered in order to reduce the frequency of recombination between homologous transgenes and infecting viral genomes. One consists in eliminating any terminal sequences derived from the donor viral genome from the transgene. The incorporation of transgene sequence into the

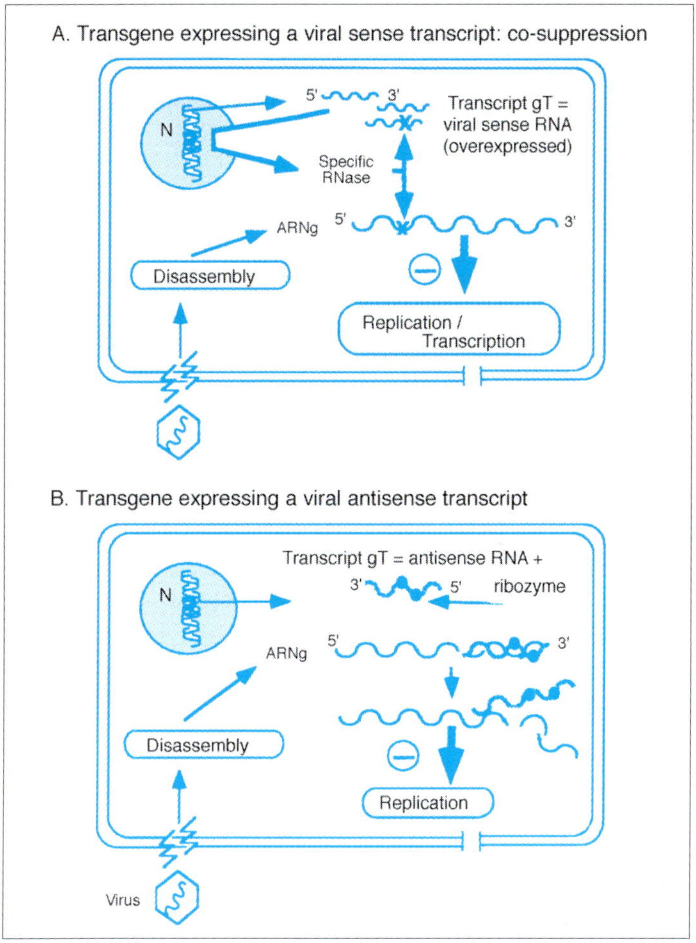

Figure 2. Resistance induced by a non-coding viral transgene.

A. Good resistance may be observed when the plant abundantly expresses inactive (untranslated) transcripts of the same polarity as the viral genome. The mechanism suggested in the case of potyviruses relies on the activation of a system of cellular ribonucleases able to specifically degrade the overexpressed transcripts as well as the gRNA in the region homologous to the transcript. This halts the viral replication. This phenomenom, called co-suppression, is due to the induction of an active but specific type of defence in the infected plant. **B.** The transgene produces RNA molecules complementary to a region of viral genomic RNA. Ribozyme sequences are included in the transgene RNA in the example given. The pairing of these transcripts with the gRNA cleaves the latter at specific points. Here too, this inhibits replication.

virus may then only occur *via* double recombination. A second precaution, applicable if the resistance is effectively conferred by the protein product and not the RNA, involves the use of a transgene modified by permutation of codons or insertion of frame-shifting spacers. Finally, in the ribozyme approach already mentioned, the probability of recombination is almost eliminated due to the short length of viral sequences required for ribozyme/viral RNA recognition. However, the weak point in this approach is the relatively high risk of escapes (mutants with mutations rendering the ribozyme sequences inoperative).

Resistance by Expression of a Co-Replicative Transcript

In certain experimental approaches, the induced resistance relies on the amplification of the transcript. This will directly or indirectly interfere with the replication of the infecting genome. The transgenes used code for satellite RNA or a defective form of the viral genome [4]. Plant satellite RNAs are molecules that are only able to multiply and reproduce in the presence of a specific helper virus. These small RNAs are, in principle, non-coding. They do not have perceptible sequence homology with the helper virus that they depend on for replication and encapsidation [19]. About thirty virus/satellite RNA pairs have already been detected. Some of them have the particularity of interfering with the multiplication of the helper virus and attenuating the severity of the disease symptoms. However, the mechanisms are still not clearly understood. The use of satellite RNA as a means of biological control has already been tested in the field in several countries without any harmful effects [20]. Tests in a confined environment have established that a satellite RNA, expressed from a transgene this time, may also attenuate the disease caused by the helper virus [21].

The specific risk of this type of approach is directly related to the amplification of the transcript triggered during the infection of the transgenic plant by the relevant virus. The attenuation of the symptoms seems closely related to this. A specific feature of RNA-dependent RNA polymerases is the absence of error correction. The rate of mutation observed in the RNA viruses is relatively high, from one to several mutations/genome/cycle [22]. The potential impact of such variability has to be considered in the case of satellite RNA. Several isolates are known to cause necrosis in some of their host plants. The nucleotide sequences of these isolates may be very similar to those of asymptomatic ones [23]. However, genetic drift of a satellite RNA sequence is not likely to cause problems unless it is associated with a selective advantage so that it spreads within the population. The factors affecting the composition of a population of satellite RNAs are still poorly understood and may vary according to the host plant [24].

Possible improvements would be to modify the transgenes to prevent the encapsidation of defective or satellite RNA transcripts or to irreversibly inactivate the sequences involved in causing necrosis.

Prospects Related to the Introduction of Virus-resistant Transgenic Plants in Farming Practice

The agronomic potential of virus-resistant transgenic plants is highly promising. There are several benefits expected from the introduction of such plants in agriculture:
- reduction in the use of pesticides to control populations of insects, nematodes and soil fungi that act as vectors for the propagation of viral diseases;
- replacement of the practice of voluntary infection of a crop with a hypovirulent strain that may reduce the yield;
- significant improvement in yield.

The risks potentially induced by the introduction of such plants, as long as the expressed transgenes do not induce a pleiotropic effect on the phenotype of the transformant and do not conceal any potential danger, are limited to the genetic dissemination of the transgene on the one hand and on the genotypic or phenotypic evolution of the infecting viral populations on the other hand.

The question of the dissemination of the transgene and the harmful effects that may result mainly involves plants (such as rape or beet) that are able to produce fertile hybrids by cross-fertilization with adventitious plants. This goes well beyond the topic of virus-resistant plants and is discussed elsewhere in this book. The introgression of a gene conferring resistance to a pathogen in the genome of a plant considered to be a weed is a plausible scenario. The potential impact should be assessed on a case-by-case basis, not only according to the biology (distribution, anatomy) of the species involved but also to the type of transgene. It is likely that the dissemination, when expected, will be all the more effective if the conferred resistance is complete and impossible to overcome. Moreover, a specific aspect that has to be taken into account in the case of virus-resistant plants is that the selection pressure that determines the biological advantage cannot be suspended or modulated, as it can with a herbicide for example. For this reason, a return to a past situation may be difficult.

The potential impact of virus-resistant plants on the evolution of infecting viral populations depends on several factors: heteroencapsidation, recombination and mutation. The molecular mechanisms involved have been described earlier. The question is whether the large-scale cultivation of these plants will increase the frequency of these events and, if so, will this give rise to viruses with increased virulence or that are more difficult to control than those derived from the spontaneous evolution of currently circulating viral populations [7,25,26]. The epidemiological assessment of the potential frequency of the phenomena mentioned above should take the following elements into account:

- Besides certain specific and possibly identifiable situations, the risks related to the heteroencapsidation of an infecting genome by a capsid protein derived from a genetically distant virus are low, and in practice non-existent beyond 30% difference in the polypeptide sequences. The possible incidence of "legitimate" heteroencapsidation (between genetically similar donor and target viruses) should be assessed according to the relative level of expression of the protein from the transgene compared to that during a viral infection, and also according to the epidemiological context. If a harmful effect does occur, the situation can be reversed since the genetic structure of the viral population involved is not affected.

- Recombination between transgene transcripts and infecting genomes is only likely to be frequent when very closely related sequences are involved (more than 90% identical). A reasonable approach would be to assess the risk related to the emergence of recombinants on the basis of the frequency of homologous (or almost homologous) intergenomic recombination in the virus family from which the transgene was taken. This can be measured *in vitro* or better still *in planta* without selection pressure.

- Different technical solutions may be considered for optimizing the biological confinement of transgene products, such as the modification of the capsid genes in order to prevent encapsidation or vector-mediated transmission and the elimination of the sequences involved in necrosis from satellite RNA. The limitation of the potential transmissibility of co-amplified transcripts would also increase safety when using such a transgene. Recombination between transgene transcripts and infecting genomes is, however, much less controllable, although still not indicative of danger.

- The potential phenotypic modifications in a virus infecting a plant that expresses an insufficiently confined transgene may include changes in virulence, transmission and, more rarely, host range. In fact, adaptation to a host or a set of hosts is often multigenic in plant viruses [27]. The ability to infect a species in a separate taxonomic group therefore requires the simultaneous adaptation of several genes or even non-coding regions. However, most of the transgenes used are only able to lead to the complementation of a single gene (recombination) or its product (heteroencapsidation).

- In general, the risks involved in creating virus-resistant plants are not specifically inherent to the use of genetic transformation. The simultaneous infection of a plant by several more or less related viruses (likely to induce heteroencapsidation and recombination), mutagenesis and the emergence of necrosis-inducing satellite RNAs are all possible independently of the introduction of transgenic plants. It is not certain that the use of such plants will significantly increase the considerable variability noted in circulating viral populations. The question of the relative speed of the mutagenesis of viral populations harboured by wild, naturally resistant plants is unfortunately still not documented. In any case, the recurring problem of viruses overcoming resistance, already observed with non-transgenic plants, does not *a priori* pose novel difficulties with transgenic plants.

- Nevertheless, certain specific situations involving transgenic plants may take on a new character and which has to be examined. By way of example, the expression of a transgene in normally uninfectable tissue or in a plant that is not a host of the donor virus, the strong and permanent expression of a transcript, and perhaps also the induction of a durably tolerated infectious state in perennial species, are all elements that generate new interactions or are likely to increase their frequency.

It does not seem wise, on the basis of the current data and considering the diversity of the plant/virus pairs that are candidates for such an approach, to recommend one or more strategies as being particularly safe. In principle, the expression of non-coding, non-amplifiable transcripts offers the best guarantee of safety. That said, the solidity of the resistance is also a safety criterion as long as it results in the marked or complete inhibition of the multiplication of the infecting virus, thereby limiting interactions inducing risk. At the same time, it is necessary to consider that a gene conferring strong and stable resistance may be invasive and hence one needs to assess the eventual consequences on the ecosystem.

By way of conclusion, current knowledge indicates that, in general, the introduction of virus-resistant transgenic plants should only involve limited risks. However, only further tests associated with monitoring of the critical stages of development will provide the necessary time for assessment. In this context, it seems wise to:

- continue the tests to increase knowledge about the determinism of the resistance and the structure-function relationships of the viral components involved, to improve the safety of the transgenes used and better assess the frequency of potentially risk-inducing biological effects;
- carry out a specific assessment of each host/transgene/virus system, taking into account the diversity of the epidemiology and the applicable strategies. Although the diversity of the latter is an asset in optimizing biosafety, it renders an overall assessment of their safety useless;
- make sure that an epidemic-monitoring system, when relevant, be set up during the introduction of a new virus-resistant variety.

References

1. Whitham S, Dinesh-Kumar SP, Choi D, Hehl R, Corr C, Baker B. The product of the tobacco mosaic virus resistance gene N: similarity to toll and the interleukin-1 receptor. *Cell* 1994 ; 78 : 1101-15.
2. Zaccomer B, Haenni AL, Macaya G. The remarkable variety of plant RNA virus genomes. *J Gen Virol* 1995 ; 76 : 231-47.
3. Fichten JH, Beachy RN. Genetically engineered protection against viruses in transgenic plants. *Annu Rev Microbiol* 1995 ; 76 : 231-47
4. Wilson TMA. Strategies to protect crop plants against viruses: pathogen-derived resistance blossoms. *Proc Natl Acad Sci USA* 1993 ; 90 : 3134-41.
5. Hackland AF, Rybicki EP, Thomson JA. Coat protein-mediated resistance in transgenic plants. *Arch Virol* 1994 ; 139 : 1-22.

6. Lecoq H, Ravelonandro M, Wipf-Scheibel C, Monsion M, Raccah B, Dunez J. Aphid transmission of a non-aphid-transmissible strain of Zucchini yellow mosaic potyvirus from transgenic plants expressing the capsid protein of plum pox potyvirus. *Mol Plant Microb Interact* 1993 ; 6 : 403-6.
7. Tepfer M. Viral genes and transgenic plants: what are the potential environmental risks? *Bio/Technology* 1993 ; 11 : 1125-32.
8. Kawchuk LM, Martin RR, McPherson J. Resistance in transgenic potato expressing the potato leafroll virus in Russet Burbank potato plants. *Mol Plant Microb Interact* 1990 ; 3 : 301-7.
9. Longstaff M, Brigneti G, Boccard F, Chapman S, Baulcombe D. Extreme resistance to potato virus X infection in plants expressing a modified component of the putative viral replicase. *EMBO J* 1993 ; 12 : 379-86.
10. Beck DL, Van Dolleweerd CJ, Lough TJ, Balmori E, Voot DM, Andersen MT, O'Brien IEW, Forster RLS. Disruption of virus movement confers broad-spectrum resistance against systemic infection by plant viruses with a triple gene block. *Proc Natl Acad Sci USA* 1994 ; 91 : 10310-4.
11. Lindbo JA, Silva-Rosales L, Proebsting WM, Dougherty WG. Induction of a highly specific antiviral state in transgenic plants: implications for regulation of gene expression and virus resistance. *Plant Cell* 1993 ; 5 : 1749-59.
12. Lai MMC. RNA recombination in animal and plant viruses. *Microbiol Rev* 1992 ; 56 : 61-79.
13. Schoelz JE, Wintermantel WM. Expansion of viral host range through complementation and recombination in transgenic plants. *Plant Cell* 1993 ; 5 : 1669-79.
14. Greene AE, Allison RF. Recombination between viral RNA and transgenic plant transcripts. *Science* 1994 ; 263 : 1423-5.
15. Falk BW, Bruening G. Will transgenic crops generate new viruses and new diseases? *Science* 1994 ; 263 : 1395-6.
16. Khatchikian D, Orlich M, Rott R. Increased viral pathogenicity after insertion of a 28S ribosomal RNA sequence into the haemagglutinin gene of an influenza virus. *Nature* 1989 ; 340 : 156-7.
17. Meyers G, Tautz R, Stark R, Brownly J, Dubovi EJ, Thiel HJ. Rearrangements of viral sequences in cytopathogenic pestiviruses. *Virology* 1992 ; 191 : 368-86.
18. Kuwata S, Masuta C, Takanami Y. Reciprocal phenotype alterations between two satellites RNAs of cucumber mosaic virus. *J Gen Virol* 1991 ; 72 : 2385-9.
19. Palukaitis P, Roossinck MJ, Dietzgen RG, Francky RIB. Cucumber mosaic virus. *Adv Virus Res* 1992 ; 41 : 281-343.
20. Tien P, Wu G. Satellite RNA for the biocontrol of plant disease. *Adv Virus Res* 1991 ; 39 : 321-39.
21. Saito Y, Komari T, Masuta C, Hayashi Y, Kumashori T, Takanami Y. Cucumber mosaic virus-tolerant transgenic tomato plants expressing a satellite RNA. *Theor Appl Genet* 1992 ; 83 : 679-83.
22. Drake JW. Rates of spontaneous mutation among RNA viruses. *Proc Natl Acad Sci USA* 1993 ; 90 : 4171-5.
23. Jacquemont M,Tepfer M. Satellite RNA-mediated resistance to plant viruses: are the ecological risk well assessed? In : Khetarpal RK, Koganezawa H, Madadi A, eds. *Control of plant virus diseases*. St Paul MN United-States : APS Press, 1996.
24. Moriones E, Fraile A, Garcia-Arenal F. Host-associated selection of sequence variants from a satellite RNA of cucumber mosaic virus. *Virology* 1991 ; 184 : 465-8.
25. Hull R. The use and misuse of viruses in cloning and expression in plants. In : Fraser RSS, ed. *Recognition and responses in plant-virus interactions NATO ASI 4*. Berlin : Springer-Verlag, 1990 : 443-7.
26. De Zoeten GA. Risk assessment: do we let history repeat itself? *Phytopathology* 1991 ; 81 : 585-6.
27. De Jong W, Mise K, Ahlquist P. The multigenic nature of RNA virus adaptation to plants. *Trends Microbiol* 1994 ; 2 : 29-31.

10

Bacillus thuringiensis: an Insecticide Reservoir

Josette Chaufaux*, Vincent Sanchis, Didier Lereclus*****

Summary

This chapter describes the use of the entomopathogenic bacterium B. thuringiensis *(*Bt*) in crop protection and the development of resistance to* Bt*-based products. First, an overview of the biology of* Bt *and of the mode of action of its insecticidal toxins is presented and the following aspects of resistance to* Bt *are discussed: (1) laboratory and field selection of* Bt*-resistant insects; (2) genetics, stability and fitness cost of resistance; (3) cross-resistance and mechanisms of resistance to* Bt *δ-endotoxins. The chapter then deals with managing pest resistance to* Bt.
The insecticidal proteins (also called δ-endotoxins or Cry proteins) produced by Bt *are stomach poisons that, after proteolytic activation in the insect midgut, cause lysis of the midgut epithelial cells. The high insect specificity of these toxins has been correlated with the presence of specific receptors in the midgut of susceptible insects. Although* Bt *has been used as an effective insecticide for over 30 years, there has been little opportunity for insects to develop resistance to* Bt *in the field. There are*

* Research Engineer, INRA, Guyancourt, France.
** Senior Scientist, Institut Pasteur, Paris, France.
*** Senior Scientist, Institut Pasteur, Paris, France.

many reasons for this apparent absence of selection of Bt*-resistant insects. One is certainly that* Bt *products have not been massively used because of their relatively high cost, narrow activity spectrum, sensitivity to UV degradation and rainfall and critical timing of application that often restricts them to sophisticated pest management programmes. In contrast, several researchers have successfully selected resistance to* Bt *in laboratory experiments. Bioassays on progeny from crosses between resistant and susceptible insects and backcross data suggest that in most cases, resistance to* Bt *is primarily controlled by one or a few genes and is autosomally inherited. The most common mechanisms of resistance found seem to be receptor modification. However, in two cases, insects displaying cross-resistance to* Bt *toxins have been selected.*

The potential for the development of Bt*-resistant insect populations in the field highlights the urgent need to design programmes and develop strategies to manage or prevent this resistance from occurring. Moreover, as crops genetically engineered to express* Bt t*oxins to make them resistant to specific insect pests become a reality (the first plants containing a single insect resistance* Bt *transgene were planted commercially in the United-States in 1996) and their massive deployment becomes imminent, the question of whether the targeted insect pests will rapidly develop resistance becomes an important issue. In transgenic plants,* Bt *insecticidal proteins will be produced continually and protected from degradation. This will create strong selection pressure on insect pests and will likely result in insects rapidly building up a resistance to* Bt*. As a result, both sprays and transgenic plants would become ineffective as insect control agents.*

A number of strategies for decreasing the rate at which insects adapt to Bt *toxins produced in transgenic plants have been proposed. These strategies include: (1) engineering plants to produce* Bt *toxins only in the tissues that are prone to insect attack; (2) using crop rotation (in which transgenics may be alternated with non-transgenics), or mosaics (in which mixtures of transgenic and non-transgenic plants are grown together); (3) creating refugia (in which a portion of a field may be planted with non-transgenics); (4) developing resistance monitoring programmes. These different strategies are based on the point that genes conferring resistance to* Bt *are generally not genetically dominant. Therefore, if susceptible individuals are allowed to survive, reproduce and eventually mate with resistant insects, their offspring will be constituted of susceptible heterozygotes. If enough of these susceptible insects survive in each generation, this should strongly decrease the risk of resistant individuals taking over the population.*

The Bacterium

Bacillus thuringiensis Berliner (*Bt*) is a sporulating, Gram-positive, facultatively aerobic soil bacterium. It belongs to the Bacillacea family, including the genus *Bacillus* and *Clostridium*. It is taxonomically related to *B. cereus*. The difference between *Bt* and *B. cereus* is mainly based on whether or not a proteinaceous crystalline inclusion is synthesized during sporulation. It is often acknowledged that *Bt* is a *B. cereus* that has acquired the ability to synthesize a crystal.

Historically, *Bt* was first discovered in 1901 in Japan in a silkworm farm. It was considered to be a threat to the silk industry. In 1911, Berliner isolated and identified *Bt* in Thuringe in a population of flour moths in the attic of a mill. Following this discovery, but especially after Angus, in 1954 [1] established that the crystal was responsible for the pathogenicity, a great deal of basic and applied research was carried out on *Bt*. However, for a number of years, the *Bt* species was thought to group bacteria that were toxic only for Lepidoptera larvae. Then, in 1977 and 1983, two new strains (*israelensis* and *tenebrionis*) were isolated and characterized. They were active against dipteran and coleopteran larvae. This new insecticidal potential raised interest in the bacterium and a number of research programs were undertaken throughout the world in order to identify new strains of *Bt*. About 50,000 strains of *Bt* have been isolated from different sources (insects, soils, plants, etc.) coming from different geographic areas.

Some of these strains were identified and classified at the World Health Organization (WHO) Collaborating Centre for Entomopathogen Bacilli (Unité des Bactéries Entomopathogènes, Institut Pasteur, Paris). On the basis of their biochemical characteristics and their flagellar antigens, they were grouped into about fifty serotypes. However, this classification does not indicate the pathotype of the strain, mainly defined by proteins called δ-endotoxins that form the crystal(s) synthesized by the bacteria. One *Bt* strain generally produces several types of δ-endotoxins. The variety of these toxins determines the insecticide activity spectrum of a given strain.

Bt reproduces in a vegetative manner until the environment becomes deficient in one of the essential nutriments. The bacterium then enters a stationary phase and starts a process that leads to the formation of spores. The δ-endotoxins are produced during the stationary phase, concomitant with sporulation. They accumulate in the mother cell and form a crystal that, at the end of sporulation, may account for about 25% of the dry weight of the bacterium.

Most often, mixtures of spores and crystals are used as biopesticides. Many formulations are currently available on the market. Among the best known are Dipel® by Abbott, Biobit® by Novo and Javelin® by Sandoz. These formulations consist of natural strains of *Bt*. However, modified strains have been marketed over the last few years (Foil® and Condor® by Ecogen and MVP® by Mycogen).

These different products are effectively used in the protection of certain crops and forests as well as stored farm products. It should also be noted that the *israelensis* strain is increasingly used to control mosquito and black fly populations, whether or not they are carriers of parasitic diseases.

The δ-endotoxins, Diversity and Mode of Action

Since 1981, when the first δ-endotoxin gene was cloned [2], about one hundred δ-endotoxin genes have been isolated from different *Bt* strains. These genes are often located on conjugative plasmids [3] and are included in structures consisting of transposable elements [4]. Their diversity is, at least in part, due to these two characteristics.

The determination of the nucleotide sequence of δ-endotoxin genes is used to classify the toxins (called Cry proteins) according to their degree of similarity. *Table I* presents a classification of the best characterized δ-endotoxins. Although highly simplified (19 classes of toxins are currently distinguished), this version of the classification reflects the amazing diversity of the insecticide toxins found in the *Bt* species. This molecular diversity is fairly well correlated with the diversity of the activity spectrum of the different toxins. As indicated in *Table I*, the proteins from the same group (I, II, III) are generally active against insects belonging to the same order. However, this is not a hard-and-fast rule. In addition, there is a large variety of toxins within the same group that do not necessarily have the same insecticide activity spectrum. All CryI proteins are toxic for Lepidoptera. However, only the CryIC protein is highly toxic against certain Lepidoptera belonging to the genus *Spodoptera* [5]. CryIA proteins are not active against these insects and CryID, CryIE and CryIF proteins have an intermediate activity.

The correlation between the molecular diversity and the variability of the insecticide activity is due to the mode of action of these proteins. Once ingested by the insect, the crystals are made soluble in the intestinal tract and the δ-endotoxins, in fact protoxins, are transformed into active toxins by the intestinal proteases. In 1990, Van Rie *et al.* [6] demonstrated that the active fractions of the δ-endotoxins bind to specific receptors on the surface of the midgut epithelial cells. These authors have also demonstrated, in the Indian flour moth, *Plodia interpunctella*, that CryIA and CryIC proteins recognize different receptors.

The determination of the 3-D structure of a δ-endotoxin has revealed the presence of three distinct domains. One of the domains, consisting of β-sheets, is responsible for the specificity of the toxin for the receptor of the insect. The toxin/receptor interaction then allows another domain of the protein (consisting of α-helixes) to form a pore in the membrane and thereby disturb ion exchange which induces a modification in the intestinal pH and cell lysis. The CryIA protein receptor was recently identified in the insect *Manduca sexta*. It is a membrane aminopeptidase [7].

Table I. Classification of *B. thuringiensis* δ-endotoxins.

δ-endotoxins Class		Size (kDa)	Susceptible Insects	*B. thuringiensis* Strains (examples)	Structure of the Crystals
I	A	130-140	Lepideptora	*kurstaki*	Bipyramidal
	B			*berliner*	
	C			*entomocidus*	
	D			*aizawai*	
	E			*kenyae*	
II	A	71	Diptera and Lepidoptera	*kurstaki*	Cubic
	B	71	Lepidoptera	*kurstaki*	
III	A B	68-73	Coleoptera	*tenebriosis*	Rhombohedral
IV	A B	125-145	Diptera	*israelensis*	Spherical
V	A	81	Lepidoptera and Coleoptera	*kurstaki*	Bipyramidal
Cyt		26-28	Diptera (non specific cytolitic activity)	*israelensis*	Spherical

At the physiological level, the lysis of the epithelial cells leads to paralysis of the insect's digestive system. It then quickly stops eating. Alone, this effect of the δ-endotoxins causes the death of the insect generally 1 to 3 days after the ingestion of the crystals. However, the insect also ingests *Bt* spores along with the crystals. The change in the pH of the digestive tract, induced by the δ-endotoxins, seems to favour the germination of spores and the growth of bacteria. The result is that septicemia is almost always associated with the toxemia. It has not been fully demonstrated, although probable, that the septicemia due to the development of *Bt* spores may optimize the toxic effect of the δ-endotoxins.

The Different Aspects of Resistance to *Bacillus thuringiensis*

Resistance Observed in the Laboratory and in the Field

The first resistance to δ-endotoxins mentioned in the literature was obtained by McGaughey in 1985 in the United States [8] with the Lepidoptera *Plodia interpunctella* resistant to Dipel®, a commercial product formulated from *Bt* serotype *kurstaki*. In 15 generations, the selected population was one hundred times as resistant as the control. The insects used during this trial were directly obtained from grain silos where they usually live and where they have already been in contact with *Bt*.

Afterwards, a great many other cases of resistance to δ-endotoxins were mentioned in the literature. *Table II* sums up the resistances acquired in the laboratory and known to date.

These selections involve ten Lepidoptera species belonging to four families (*Noctuidae, Pyralidae, Plutellidae* and *Tortricidae*), two Coleoptera species (*Chrysomelidae*) and three Diptera species (*Culicidae* and *Muscidae*). Only eight of these fifteen species (five Lepidoptera, two Coleoptera and one Diptera) developed resistance exceeding ten times that of sensitive non-selected populations.

The first case of *Bt* resistance detected in nature involved a population of *Plutella xylostella*, the diamondback moth, in Hawaii. The resistance observed after more than fifty sprayings with Dipel® in the field was about 25 to 30 times [9]. Since then, Tabashnik 10 has mentioned other cases of resistance appearing in the same species in the states of Florida and New York in the United States, the Philippines, Thailand and Malaysia. In Japan, resistance of *P. xylostell*a to a commercial preparation formulated from a strain of the *kurstaki* serotype was detected in a greenhouse population. This resistance was about 700 times.

The resistances observed in the laboratory were obtained by mass selection of populations in most cases coming from the field. However, not all the attempts at selection were successful. There are several reasons for these failures: (1) the response may differ according to the species tested, (2) the *Bt* strain used may be little or not at all active on the insect tested, (3) the response to the selection within the same species is variable, it depends on the difference in sensitivity of the population samples tested, whether they come from the same geographical location or different locations.

The number of generations selected in the laboratory in order to obtain resistance is always at least equal to ten when the resistance coefficient observed exceeds ten. The only exception is *P. xylostella* where nine generations were sufficient for a 66-fold increase in resistance. However, the original population was in contact with *Bt* in the wild.

Until 1992, the resistance was thought to only involve one toxin. For example, McGaughey and Johnson [11] indicated that a strain of *P. interpunctella* resistant to the *Bt* serotype *kurstaki* remains susceptible to 21 isolates belonging to five different serotypes and producing at least one different toxin from the toxins contained in the serotype *kurstaki*. Van Rie *et al.* [6] also demonstrated that a strain of *P. interpunctella* resistant to the toxin CryIA(b), contained in the serotype

Table II. Summary of the selection of resistant insects in the laboratory with different *B. thuringiensis* δ-endotoxins (according to [10]).

Insects	Toxins or Strains of *Bt* Used	Number of Generations Selected	Selection Pressure	Acquired Resistance	Comments
Lepidoptera					
Plodia interpunctella	*Bt kurstaki*	15	50	100	Insects obtained from grain silos, having been in contact with *Bt* before selection
Cadra cautella	*Bt kurstaki*	21	70-90	7	Low resistance
Anagasta kühniella	*Bt "Anduze"*	7	50	1	No resistance
Homosoma electum	*Bt kurstaki*	11	50	1.7	No resistance
Heliothis virescens	CryIA(b)	14	70-90	24	Selection with *P. fluorescens* expressing CryIA(b)
Spodoptera exigua	CryIC	21	70-90	100	
Spodoptera littoralis	CryIC	14	90	> 50	
Trichoplusia ni	CryIA(b)	7	70	1.4	No resistance
Plutella xylostella	*Bt kurstaki*	9	45	66	
Choristoneura fumiferana	*Bt sotto*	8		3.8	Low resistance
Coleoptera					
Chrysomela scripta	CryIIIA	24	–	> 50	
Leptinotarsa decemlineata	*Bt san diego*	12	99	60	Insects that were in contact with *Bt* in the field before selection
Diptera					
Aedes aegypti	*Bt israelensis*	14	50	1.1	No resistance
Culex quinquefasciatus	*Bt israelensis*	120	90	16	
Musca domestica	*Bt thuringiensis*	25	–	6	Insect not susceptible to the *Bt* strain used

kurstaki, was always sensitive to the toxin CryIC contained in the serotype *aizawai*. Many similar results have been published.

The first case of cross-resistance was observed by Gould *et al.* [12]. A wild population of moths, *Heliothis virescens* (tobacco hornworm), was selected for resistance to the *Bt*-toxin CryIA(c) (R = 50 after 17 generations). Cross-resistance was observed with the toxins CryIA(b) and CryIIA. More recently, Müller-Cohn *et al.* reported another case of cross-resistance [13]. It concerned *Spodoptera littoralis*, the Egyptian cotton leafworm that is resistant to CryIC (R > 500 in 14 generations). It also became resistant to toxins CryIE and CryID.

Genetic Determinism of Resistance

The genetic determinism of the acquisition of *Bt* resistance is still not clear. It can be effectively only studied with populations of insects that have become homozygous for the resistance gene(s). This does not seem to be the case for most articles dealing with resistance. For this reason, each team presents results that differ according to the species or toxins used.

The opinions also differ as regards the stability of the resistance in the absence of selection. According to Tabashnik [10], it varies according to the species and strains within the same species. The resistance of *P. interpunctella* to Dipel® (R = 60) did not decrease after 29 generations without selection [14]. Similarly, the resistance of H. virescens to the toxin CryIA(b) expressed in *Pseudomonas fluorescens* (R = 24 in 14 generations) was maintained for two generations without selection [15]. However, in the same species, acquired resistance of 69-fold dropped to 13-fold after five generations without selection pressure [16]. As regards *Leptinotarsa decemlineata*, the Colorado potato beetle, the resistance (R = 60) dropped to 3-fold after 51 generations without selection pressure [17].

Studies of the genetic determinism of *Bt* resistance also vary according to the species, toxin and experimental protocol. Marrone and McIntosh [18] estimate that the resistance to *Bt* of a population of *H. virescens* depends on several factors, that it is predominantly autosomal with several sex-related factors and it depends on the endotoxin used for the selection. McGaughey [8] and McGaughey and Beeman [14] found that the resistance to *Bt* of a population of *P. interpunctella* with past exposure to Dipel® in the wild is more or less recessive and suggest that it depends on a single factor. In another study on the resistance of *H. virescens* to CryIA(b) expressed in *P. fluorescens,* Sims and Stones [16] conclude that the resistance is autosomic, incompletely dominant and depends on several genetic factors. This gives it an unstable character. Finally, in an article on *P. xylostella* that became resistant to a commercial formulation of *Bt*, Hama *et al.* [19] conclude that the resistance is related to one or very few recessive genes. This is the conclusion of most authors.

Table III provides a summary of the knowledge acquired concerning the genetics and stability of *Bt* resistance in the four most studied species to date.

Table III. Summary of the knowledge concerning genetics and the stability of the resistance of insects to *B. thuringiensis*.

Insects	Toxins or Strains of *Bt* Used	Selection Pressure and Resistance Observed	Stability of the Resistance	Resistance Genetics
Plodia interpunctella	Dipel® containing: CryIA(b), CryIA(c), CryIIA and CryIIB	low: 150 high: > 250	stable: no reduction in R after 21 generations without selection	partially recessive autosomic, monogenic
Plutella xylostella	Dipel®	field + laboratory R: from 1800 to 6300	reversible when the selection pressure was interrupted	not determined
Heliothis virescens	CryIA(c), CryIA(b) then Dipel®	R: 57 (Dipel®) R: 69 (CryIA(b)) R: 16 (CryIA(c))	R decreased from 69 to 13 after 5 generations without selection pressure	autosomic, partially dominant for CryIA(b), partially recessive for CryIA(c), probably polygenic for CryIA(b)
Leptinotarsa decemlineata	*Bt tenebrionis* containing CryIIIA	R: 60	declined by a factor of 3 after 51 generations without selection pressure	autosomic, partially dominant

Little information is available concerning the biological cost of the acquisition of *Bt* resistance in insects. In a Dipel®-resistant strain of *P. xylostella*, the fertility and rate of reproduction are significantly inferior to that of a susceptible strain, although no significant differences were noted in the duration of larval development, the percentage survival from egg to adult (in the absence of any treatment) and the weight of the chrysalis. However, no significant differences were found in the percentage survival and larval weight between a sensitive and a resistant strain of *H. virescens*. In a CryIC-resistant *S. littoralis* population, no differences were noted compared with a sensitive population as regards the weight of the chrysalis, the duration of larval development or the sex-ratio during selection.

All of these results indicate great variability in the response to selection according to the insects tested and the toxins used. Each case seems to be specific and requires individual study.

Molecular and Cellular Mechanisms of Resistance

Biochemical Factors determining Resistance to Insecticides

The efficiency of an insecticide depends, to a large degree, on the quantity of active ingredient that reaches the molecular target. It is modified by the different factors which, in insects, help limit or reduce the quantities of insecticide found in the insect. Among these factors, it is important to note the following:
- the speed of penetration of the insecticide (that may vary from one insect to the next),
- the number and/or quantity of binding or detoxification proteins likely to trap and/or metabolize the insecticide,
- the speed of elimination (excretion) of the insecticide before or after being metabolized (or more generally the time that the active form of the insecticide remains in the insect),
- the number and affinity of the target molecules (receptors) found in the insect.

The proteins involved in each of these stages are most often found both in the sensitive insects and those that have acquired a certain level of resistance. In most cases, a qualitative or quantitative modification in one or several of these proteins renders certain insects less susceptible (resistant) to the insecticide. The resistance is therefore a pre-adaptation phenomenon and the selection of the least sensitive individuals is only the result of the exposure of a population to an insecticide. In other terms, this means that the resistant individuals are already found in the population and that natural selection only favours the emergence of the genotypes that are best adapted to the new environmental situation created by the use of the insecticide. However, the mechanism of resistance involved is always closely related to the specific mode of action of each insecticide and/or the physiology of the organ or target molecule.

δ-endotoxins of **B. thuringiensis** *and their Cellular Target*

As regards the δ-endotoxins of *Bt*, one of the first steps in the intoxication process is the fixation of the toxin on the specific receptors located at the surface of the gut cells. The host spectrum of these toxins is mainly determined by the presence of these specific receptors in the target insect. The phase following the binding of the toxin on the receptor and the eventual contribution of the receptor to the toxicity have not yet been determined. However, it is generally agreed that the toxin acts by osmo-colloidal cytolysis following the formation of pores in the gut cells [20]. The δ-endotoxins may thereby be classified among the bacterial toxins that do not need to be internalized to be active. Moreover, the δ-endotoxin receptors are located at the surface of the microvilli (or brush border) on the apical membrane of the midgut cells in larvae. These cells, the site of most absorption and digestion of ingested food, are separated from the lumen of the digestive tract by a peritrophic membrane that acts like a filter surrounding the gastric contents and preventing imperfectly solubilized particles from coming into contact with the apical microvilli. In certain Lepidoptera, it can only be crossed by particles or polypeptides not exceeding 7 or 8 nm in diameter.

Resistance to B. thuringiensis

In view of the above, the solubilization of the ingested crystals and the activation of protoxin molecules by proteolysis are the two steps required to give δ-endotoxins their insecticidal activity. One of the possible mechanisms for δ-endotoxin resistance may involve a change in the pH or the prevailing reducing capacity in the gut of larvae, preventing the dissolution of the crystals. They would then simply be eliminated by passage through the intestine. A variation in the type, quantity or specificity of the proteolytic enzymes that alter the activation of the δ-endotoxins is also possible. The molecular weight of a protoxin molecule that is not activated or imperfectly activated would be too high and not allow it to cross the peritrophic membrane. In fact, a resistance mechanism of this type was recently demonstrated in a population of *P. interpunctella* [21]. In this population of insects, the resistance seems to be related to faulty protoxin activation by the insect proteases.

Another essential step in the action mechanism of δ-endotoxins involves the fixation of the activated toxin on the specific receptors. Until 1992, all the articles published on the resistance of insects to *Bt* indicate selective resistance with respect to different δ-endotoxins. For this reason, a resistance mechanism involving modification of the specific receptor of each δ-endotoxin was proposed and investigated in several of the cases of resistance described. Experiments involving the fixation of radio-labelled toxins on bladders prepared from brush border cells, obtained from intestines of susceptible or resistant insects, have demonstrated that, in most cases, the resistance was due to a modification in the number or affinity of the toxin receptors [6].

However, Gould *et al.* [12] have demonstrated that cross-resistance to several *Bt* toxins is possible without a modification in the number of receptors or their affinity for these toxins. This suggests that unspecific resistance mechanisms are also possible. One can also imagine resistance mechanisms that intervene in the stages following the fixation of the toxins on their receptors (insertion of the toxin in the epithelial cells and pore formation). Such possibilities have not yet been explored since the tools required for this type of study are still not available. Finally, a resistance mechanism by modification of the behaviour or habits of the insect is also possible. This has been searched for, without success, in a case of *Bt* resistance.

Ecological Risks Related to the Use of Transgenic Plants Expressing δ-endotoxins

The Development of Resistant Insects and the Possible Consequences

Bt preparations have been authorized throughout the world in order to control certain agricultural and forest pests (since the 60's in the United States, 1970 in France). The first case of resistance in an agricultural situation was noted in 1990 in Hawaii. Commercial products are made from spore and crystal preparations containing several toxins. For a susceptible species, this reduces the probability of acquired resistance. Since the first experiment describing the transfer of a *Bt* gene to tobacco [22], many other plants (for example, rice, cotton, potato, maize) have been transformed. Some of these plants express δ-endotoxins at a sufficient level to be protected against the damage caused by the insects. However, since first generation transgenic plants only synthesize one toxin, the risk of selecting a resistant population of insects is probably higher. In addition, the δ-endotoxins produced by transgenic plants may be broken down less quickly than those sprayed during standard treatments. For this reason, they may be considered to be more persistent, thereby increasing the selection pressure and the risk of selecting resistant insects more quickly.

The main ecological implication related to the use of transgenic plants involves the development and speed of the appearance of insects resistant to *Bt* δ-endotoxins.

The development of resistance to a *Bt* toxin in major pests would render ineffective the classic treatments carried out with a biopesticide containing the same toxin. The main ecological effect of the acquisition of this resistance may thereby be an increase in the density of the population of this pest. This would result in major damage to some crops. This would lead to an increase in the use of other methods, mainly chemical, with the undesirable environmental effects that we are familiar with.

However, it is also possible that the acquisition of resistance by a pest population may have a biological cost that makes it more vulnerable to parasites and diseases. In this case, other biological control methods may still be possible. This does not exclude chemical control. Cross-resistance with other chemical insecticides has been noted to date due to the way *Bt* resistance is acquired.

A second ecological implication related to the use of transgenic plants involves the risk of the transfer of δ-endotoxin genes to other wild or cultivated species. If the genes coding for the *Bt* toxins are introduced into cultivated plants that are genetically close to other native wild species, the risk of dissemination of δ-endotoxin genes (in particular by the pollen) is not negligible. The question is then to determine whether the acquisition of an insect resistance gene by a natural species may increase the selection pressure, in particular in the case of polyphagous pests, and thereby increase the speed of appearance of resistant insects.

However, it should be noted that the three transgenic plants containing δ-endotoxin genes from *Bt* that are about to be commercialized (cotton, potato and maize) do not have close relatives in most of the zones where they are currently grown. This reduces the risk of gene dissemination to wild species.

In the case of transgenic plants that are interfertile with adventitious species, a network should be set up to monitor insect populations developing a resistance to *Bt* toxin, especially in the case of polyphagous insects. Here too, any intervention using a biopesticide containing *Bt* would become ineffective and this would harm the crops involved.

The Case of Transgenic Maize

Maize expressing the *Bt* toxin CryIA(b) is resistant to the main pest attacking this crop: *Ostrinia nubilalis*, the European corn borer. The strategy of using transgenic plants provides several advantages. The pest, due to its endophytic habit, is difficult to reach with a standard insecticide treatment. After hatching, the caterpillars immediately feed on leaves in which the toxin is synthesized and thus rapidly come into contact with the insecticide. They are neutralized before causing major damage. With this technique, the treatment does not have to be repeated. In addition, the toxin, sensitive to ultraviolet light, is protected from unfavourable climatic conditions.

The risk of finding a population of moths that are resistant to the toxin synthesized by maize is, according to the current state of knowledge, difficult to estimate. Depending on the region, the European corn borer goes through one to three generations per year. In the cases described to date, the resistance appears when the pests undergo a strong selection pressure for at least ten generations, in the absence of genetic exchanges with susceptible populations.

Due to this potential risk, a comparative study of the spatial and temporal evolution of the susceptibility of different pests is recommended. This may be carried out by establishing toxicology curves using biological assays on insects, *i.e.* noting the evolution of the susceptibility of a pest to a *Bt* toxin in the course of time.

It is also possible to monitor the spatial and temporal expression of the toxin during the growth of the plant, in parallel with the mortality of the target insects. This approach would reveal the real impact of the toxin on the pests and whether or not the surviving insects were in contact with the toxin. It would then be possible to verify whether the susceptible individuals remain on the site, thereby delaying the appearance of a resistant homozygous population.

Recommended Strategies for Delaying the Appearance of Resistance

The different strategies developed to delay the appearance of resistance in the field are still only assumptions, especially those that involve the use of genetically modified plants. Their validation in the laboratory and then in the field should confirm their reliability.

As regards the pests treated by classic sprays, the use of insecticides with different modes of action is recommended either with a very short interval (five to seven days) or at each generation. This is the most effective solution according to Marrone and McIntosh [18].

Several strategies can be considered in the case of transgenic plants. Many researchers are considering the construction of plants expressing several *Bt* toxin genes [23] or genes from different sources, like certain protease inhibitors, conferring a wider activity spectrum. This approach has two advantages: it may delay the appearance of resistance to toxins binding to different receptors and it may reveal a possible synergy.

The use of promoters induced by a xenobiotic or by a plant wound (management of the expression of the toxin over time) and the production of plants where the toxin is synthesized in a tissue-specific manner (management of the expression of the toxin in space) are also being studied.

The creation of refuge zones at different levels – among the different parts of the same plant, among plants in the same field or between fields by planting mixed crops – enables the dilution of the resistant survivors by crossing them with insects that have not been in contact with the toxin. This would delay the appearance of homozygous RR populations. It is also possible to consider crop rotations with other plants or with non-transgenic plants of the same species.

Conclusion

So much progress has been made in understanding the mode of action and genetics of *Bt* over the last ten years that a great many strategies have been developed. However, the future use of transgenic plants expressing *Bt* toxin genes raises certain ecological problems. Considering the diversity of the δ-endotoxin genes,

the multitude of geographic locations likely to be involved and the many different plant-based technologies, it is advisable to begin research programs to establish the scientific basis for the management of transgenic plants in four areas:
- the continuing study of δ-endotoxin receptors;
- the genetic basis and heredity of *Bt* resistance;
- the methodologies for assessing the risk of the appearance of resistance in populations of target pests;
- the evaluation of resistance management strategies in the field (currently these are still all theoretical).

References

1. Angus TA. A bacterial toxin paralysing silkworm larvae. *Nature* 1954 ; 173 : 54-6.
2. Schnepf HE, Whiteley HR. Cloning and expression of a *Bacillus thuringiensis* crystal protein gene in *Escherichia coli*. *Proc Natl Acad Sci USA* 1981 ; 78 : 2893-7.
3. Gonzàlez JM, Brown BJ, Carlton BC. Transfer of *Bacillus thuringiensis* plasmids coding for delta-endotoxin among strains of *B. thuringiensis* and *B. cereus*. *Proc Natl Acad Sci USA* 1982 ; 79 : 6951-5.
4. Lereclus D, Ribier J, Klier A, Menou G, Lecadet MM. A transposon-like structure related to the δ-endotoxin gene of *Bacillus thuringiensis*. *EMBO J* 1984 ; 3 : 2561-7.
5. Sanchis V, Lereclus D, Menou G, Chaufaux J, Lecadet MM. Multiplicity of δ-endotoxin genes with different specificities in *Bacillus thuringiensis aizawa* 7.29. *Mol Microbiol* 1988 ; 2 : 393-404.
6. Van Rie J, McGaughey WH, Johnson DE, Barnett BD, Van Mellaert H. Mechanism of insect resistance to the microbial insecticide *Bacillus thuringiensis*. *Science* 1990 ; 247 : 72-4.
7. Knight PJK, Crickmore N, Ellar DJ. The receptor for *Bacillus thuringiensis* CryIA(c) delta-endotoxin in the brush border membrane of the lepidopteran *Manduca sexta* is aminopeptidase N. *Mol Microbiol* 1994 ; 11 : 429-36.
8. McGaughey WH. Insect resistance to the biological insecticide *Bacillus thuringiensis*. *Science* 1985 ; 229 : 193-5.
9. Tabashnik BE, Cushing NL, Finson N, Johnson MW. Field development of resistance to *Bacillus thuringiensis* in diamondback moth (Lepidoptera: Plutellidae). *J Econ Entomol* 1990 ; 83 : 1671-6.
10. Tabashnik BE. Evolution of resistance to *Bacillus thuringiensis*. *Ann Rev Entomol* 1994 ; 39 : 47-79.
11. McGaughey WM, Johnson DE. Indianmeal moth (Lepidoptera: Pyralidae) resistance to different strains and mixtures of *Bacillus thuringiensis*. *J Econ Entomol* 1987 ; 85 : 1594-600.
12. Gould F, Martinez-Ramirez A, Anderson A, Ferré J, Silva FJ, Moar WJ. Broad-spectrum resistance to *Bacillus thuringiensis* toxins in *Heliothis virescens*. *Proc Natl Acad Sci USA* 1992 ; 89 : 7986-90.
13. Müller-Cohn J, Chaufaux J, Buisson C, Gilois N, Sanchis V, Lereclus D. *Spodoptera littoralis* (*Lepidoptera: Noctuidae*) resistance to CryIC and cross-resistance to other *Bacillus thuringiensis* crystal toxins. *J Econ Entomol* 1996 ; 89 : 791-7.
14. McGaughey WM, Beeman RW. Resistance to *Bacillus thuringiensis* in colonies of indianmeal moth and almond moth (*Lepidoptera: Pyralidae*). *J Econ Entomol* 1988 ; 81 : 28-33.
15. Stone TB, Sims SR, Marrone PG. Selection of tobacco budworm for resistance to a genetically *Pseudomonas fluorescens* containing the δ-endotoxin of *Bacillus thuringiensis* subsp. *kurstaki*. *J Invert Pathol* 1989 ; 53 : 228-34.

16. Sims SR, Sone TB. Genetic basis of tobacco budworm resistance to an engineered *Pseudomonas fluorescens* expressing the δ-endotoxin of *Bacillus thuringiensis. J Invert Pathol* 1991 ; 57 : 206-10.
17. Whalon ME, Miller DL, Hollingworth RM, Grafius EJ, Miller JR. Selection of a Colorado potato beetle (*Coleoptera: Chrysomelidae*) strain resistant to *Bacillus thuringiensis. J Econ Entomol* 1993 ; 86 : 226-33.
18. Marrone PG, McIntosh SC. Resistance to *Bacillus thuringiensis* and resistance management. In : Entwistle PF, Cory JS, Bailey MJ, Higgs S, eds. *Bacillus thuringiensis, an environmental biopesticide: theory and practice*. Chistester, New York, Brisbane, Toronto, Singapore : J. Wiley & Sons, 1193 : 221-35.
19. Hama H, Suzuki K, Tanaka H. Inheritance and stability of resistance to *Bacillus thuringiensis* formulations of the diamondback moth, *Plutella xylostella Linneaus* (*Lepidoptera: Yponomeutidae*). *Appl Entomol Zool* 1992 ; 27 : 355-62.
20. Knowles BH, Dow JAT. The crystal δ-endotoxins of *Bacillus thuringiensis*: models for their mechanism of action on the insect gut. *BioEssays* 1993 ; 15 : 469-76.
21. Oppert B, Kramer K, Johnson DE, MacIntosh SE, Mc Gaughey WH. Altered protoxin activation by midgut enzymes from a *Bacillus thuringiensis* resistant strain of *Plodia interpunctella. Biochem Biophys Res Commun* 1994 ; 193 : 940-7.
22. Vaeck M, Reynaerts A, Höfte H, Jansens S, De Beukeleer M, Dean C, Zabeau M, Van Montagu M, Leemans J. Transgenic plants protected from insect attack. *Nature* 1987 ; 328 : 33-7.
23. Van der Salm T, Bosch D, Honée G, Feng L, Munsterman E, Bakker P, Stiekema WJ, Visser B. Insect resistance of transgenic plants that express modified *Bacillus thuringiensis* CryIA(b) and CryIC genes: a resistance management strategy. *Plant Mol Biol* 1994 ; 26 : 51-9.

Conclusion

The French Biomolecular Engineering Commission, Scientific Data and the Public Debate

Axel Kahn*

During the ten years since the French Biomolecular Engineering Commission was created, the economic prospects for genetic engineering in agriculture and the food industry have changed, as has public opinion. Today, in spite of brilliant economic perspectives, the public acceptance of this type of technique remains low in Northern and Eastern Europe and has recently decreased in France and Southern Europe, mainly due to the bovine spongiform encephalitis outbreak. However, there is obviously no relationship between modern biotechnology and the circumstances of the spread of this disease. Within the current context, the Commission wanted to present, as objectively as possible, the scientific and technical data associated with the large-scale farming and commercialization of transgenic plants, and to present the fruit of these 10 years of evaluation of a large number of field tests on transgenic plants.

The French Biomolecular Engineering Commission (in French: Commission du Génie Biomoléculaire, CGB) is ten years old. During this period, the scientific and technical context has changed and so has public opinion.

* Ex-President of the Biomolecular Engineering Commission, Paris, France.

In 1987, when the first favourable decisions were made regarding experiments on transgenic plants in the field, the goals were only experimental, the economic prospects uncertain and there was practically no public interest, at least in France.

After Mr. François Guillaume, Minister of Agriculture at that time, created the CGB in 1986, a media event was organized in collaboration with a major industrialist and in the presence of the Ministers of Agriculture and the Environment (Mr. Brice Lalonde), the members of the CGB and a large number of representatives from the world of agriculture and the phytosanitary and food industries. I remember the perplexity of the journalists present. They wondered whether these questions involving transgenic plants and, more generally, the deliberate release of Genetically Modified Organisms would interest their readers.

Ten years later, Mr. Le Buanec summed up the available data concerning the economic prospects of genetic engineering in agriculture. They are considerable and probably even exceed those of the drug industry. It isn't by chance that the agro-food industry and nutrition have been the focus of major industrial mergers during the last year. In the biotechnologies, we note the creation of Novartis resulting from the merger of Ciba-Geigy and Sandoz and the restructuring of Agrevo within the Hoechst Group, associated with the purchase of PGS (Plant Genetic System), etc. only to mention Europe.

In the United States, Canada and China, large-scale farming of transgenic plants has begun. The number of commercialized varieties is quickly increasing. Therefore, by means of international trade, plant products derived from transgenic plants will obviously be available on the world market. At the same time, the indifference of French and European public opinion has turned into anxiety, if not hostility.

The prospects for this field of activity were disturbed by a combination of several phenomena. Although they are not related to genetic engineering in agriculture, they involve human food and the symbolic value remains extremely strong. The most evident disaster involves bovine spongiform encephalitis. The public saw it as the result of the artificial manipulation of the methods used to prepare human food. In the unconscious, this remains deeply associated with the idea of "natural". In this way, cattle feed made of animal meal obtained from the bodies of dead animals was confused with other human activities that seem to modify traditional methods of food preparation. For this reason, transgenic plants were lumped in consumers' repugnance together with the image of cows that are no longer allowed to graze but that have become carnivorous and even cannibals. Naturally, only indignation can hold this unlikely amalgam together. In fact, conferring a given character to a plant using a specific and perfectly characterized gene differs greatly from feeding animals with an uncertain product obtained from a mixture of dead bodies of cows, sheep and goats. It is possible to argue that the selectivity of genetic engineering is one possible answer to the uncertainties associated with the use of complex products. To take a good example concerning human health, one regrets that the anti-haemophilia factor produced by genetic engineering was not available before tens of thousands of haemophiliacs throughout the world were contaminated with the AIDS virus found in the fractions prepared from human plasma.

This receptivity of public opinion to vengeful talk about adulterated foods prepared by perverse biotechnicians and industrialists was naturally exploited by citizens' groups who consider genetic engineering to be intrinsically condemnable. This is certainly a good strategy in a democratic country, especially as it has replaced efforts that have been fruitless in the past. In fact, several years ago, major campaigns were waged against drugs produced by genetic engineering and the factories producing them.

Nobody today questions that the human insulin produced by genetic engineering represents progress, that the vaccine against hepatitis B, also produced by genetic engineering, can avoid the development of a large number of liver cancers in the world and that, as noted above, recombinant growth hormone or anti-haemophilia factors available today would have avoided the iatrogenic transmission of Creutzfeldt-Jacob's disease and AIDS. However, genetic engineering in agriculture remains controversial and the debate is still raging today.

But the power of genetic engineering to improve and diversify cultivated plants justifies that the remaining uncertainties concerning certain strategies and products be dealt with. We have tried to do this in this book by examining the problems associated with gene flow, the dissemination of herbicide resistance genes and the possible appearance of resistance in pathogens to the defence mechanisms introduced into transgenic plants.

The current situation is as follows: certain strategies and certain types of transgenic plants have been tested in the field for five to ten years. Within this context, no immediate or potentially unfavourable phenomena have been noted. However, the limited scale of the tests indicates that these conclusions have to be considered as temporary. The projects that seem to be potentially dangerous have been eliminated. For other projects, the theoretical possibility of unfavourable consequences exists when the plants are cultivated on a much bigger scale. If such consequences appear to be serious and uncontrollable, even though they may never have been observed, it would be wise (in accordance with the principle of reasonable precaution) to eliminate the risk, for example by forbidding the transgenic plants concerned. However, the potential dangers are most often relatively benign and, in any case, easily controlled and reversible if anything untoward happens.

In this context, there are two opposing positions nowadays. The first proposes a moratorium, in reality a prolongation of the period of uncertainty. The second proposes the temporary and conditional commercialization of these products, along with the monitoring of the consequences of the large-scale farming of these plants. The CGB supports the second position. In any case, the companies that put products on the market should be required to specify how they will counter any undesirable consequences should they arise. After several years (3 to 5 years), the analysis of the data collected will confirm the commercialization in certain cases and revoke it in others. It will be up to the companies concerned to eliminate the organisms that are the source of the undesirable effects observed. We feel that this is the only attitude which will both further our knowledge about these new techniques and allow us to reap the expected benefits from them. This seems to be the best way to take into account the economic interests involved, public health requirements and the need for an adequate and continuous supply of food in the world.

Achevé d'imprimer par Corlet, Imprimeur, S.A. - 14110 Condé-sur-Noireau (France)
N° d'Imprimeur : 37474 - Dépôt légal : juin 1999 - *Imprimé en U.E.*